NONRANDOM

NONRANDOM

The Secret Life of DNA Mutations

NOGA WIES

PELAGIC PUBLISHING

First published in 2026 by
Pelagic Publishing
20–22 Wenlock Road
London N1 7GU

www.pelagicpublishing.com

Nonrandom: The Secret Life of DNA Mutations

https://doi.org/10.53061/KHZB7219

A CIP record for this book is available from the British Library

ISBN 978-1-78427-683-6 Hardback
ISBN 978-1-78427-684-3 ePub
ISBN 978-1-78427-685-0 PDF

EU Authorised Representative: Easy Access System Europe – Mustamäe tee 50, 10621 Tallinn, Estonia, gpsr.requests@easproject.com

Typeset in Minion Pro by S4Carlisle Publishing Services, Chennai, India

For Daniel and Dawn

"I have steadily endeavoured to keep my mind free so as to give up any hypothesis . . . as soon as facts are shown to be opposed to it."

Charles Darwin

Contents

Preface

Mutations are commonly described as "copying errors", and their inevitability is often demonstrated by pointing out the difficulty of copying a long text by hand without making the occasional error. This metaphor, however, is woefully inadequate because it portrays mutations in an excessively random light. That is not to say that every copying error is inherently random; factors such as word length and the copier's growing boredom can predictably increase the chances of errors occurring in particular places. But generally speaking, a distracted copier is liable to err randomly throughout a given text, forcing us to look elsewhere for an apt metaphor.

How, then, can we describe mutations in a manner that truly captures their essence? Instead of focusing on the copying of written texts, perhaps we should direct our attention to the transmission of memorized works. Memorizing and repeating poems and great sagas was, after all, a practice that predated the relatively modern invention of writing. In fact, people who dedicated their lives to memorizing and reciting important religious and cultural works enjoyed a privileged status in many ancient civilizations. The Brahmins of India, for example, began memorizing and orally transmitting the Hindu *Vedas* in the second millennium BCE, and were revered as priests and spiritual teachers. So what can this ancient practice teach us about mutations?

Although the art of memorization has fallen from grace in the modern era, the American classicist Milman Parry was able to record the last vestiges of it in the early 1930s. Parry traveled to Yugoslavia, where illiterate singers called *guslari* were still reciting traditional poems – some of which were thousands of lines long – from memory. He diligently recorded the guslari reciting these epic poems as well as conversations between guslari, and later analyzed his recordings to investigate this fascinating cultural phenomenon. Interestingly, though the guslari Parry encountered were all convinced that they were repeating poems exactly as they had first heard them, recordings of two guslari reciting the same poem and even of the same guslar reciting the same poem on two separate occasions revealed that words had been swapped, lines had been rearranged, and whole segments had been inserted or left out. The poems had, in other words, *mutated* with every recitation.

The guslari's poems, as it turns out, did not mutate randomly. Every time a guslar recited a poem, it was tweaked to become *more memorable*. Thus, long words were replaced with shorter ones, plots were restructured for better flow, and uninteresting lines were left out. To be clear, the guslari were not completely (subconsciously) free to alter their poems in pursuit of memorability. Any change to a line would still need to fit the poem's meter, preserve its alliteration, and conform to additional structural constraints. Put simply, not only were the poems' mutations nonrandom, they were nonrandom on surprisingly numerous levels. Let us bear this in mind as we delve into the surreal world of genetic mutations.

CHAPTER 1

Lamarckian Lore

Evolution is an ancient idea. Back in the sixth century BCE, the Greek philosopher Anaximander of Miletus developed what is perhaps the earliest evolutionary theory. Anaximander, who may have been Pythagoras' teacher, came up with the groundbreaking notion that the universe is governed by inviolable natural laws, rendering supernatural interventions unnecessary. This might seem self-evident to twenty-first-century readers, but Anaximander's contemporaries believed that a dizzying array of supernatural beings, and the interactions between these supernatural beings as well as between them and mortals, were behind every natural phenomenon, from earthquakes to the changing seasons.

Anaximander's epiphany led him to formulate "scientific" hypotheses which, although erroneous, sound more plausible to modern ears than the mythological ideas of his time. Anaximander posited, for example, that the Earth is a short cylinder with living beings inhabiting its flat top. Unlike most of his contemporaries, Anaximander rejected the notion of the Earth resting on the shoulders or backs of Titans, turtles, or elephants. The Earth, he believed, stays in the center of the universe simply because it has nowhere to fall, and is illuminated by the hollow rings of fire we call the sun and the moon. Anaximander's evolutionary theory was a similar mixture of brilliant insights and fanciful speculations. Organisms, he realized, evolve from simpler predecessors via a process of adaptation. Anaximander also had the scientific acumen to intuit that life originated in water. Indeed, he believed that all early animal fetuses resemble fish due to their fishy origins. This hypothesis was backed by Anaximander's observations of human and fish young. Since human babies are completely helpless, he reasoned that the first babies must have developed within fishlike ancestors, and that these fishlike ancestors had been self-sufficient immediately upon hatching from eggs.[1]

Anaximander's theory of adaptive evolution was compelling, but what, exactly, could its driving force be? In the fifth century BCE, the Greek philosopher Empedocles came up with an ingenious answer – chance. Empedocles believed that life began when disembodied organs emerged from the earth. These individual organs wandered around until the power of Love joined them, combining random organs to produce hybrid creatures such as hermaphrodites and oxen with human heads. Naturally, some of the bizarre creatures struggled to survive due to their mismatched parts. These unfortunate beings quickly died out, while those blessed with complementary organs thrived and became the ancestors of modern organisms.[2]

Hippocrates, who is best known for coining the Hippocratic Oath, was a contemporary of Empedocles. Interestingly, the two great philosophers viewed evolution very differently. For Empedocles, evolution was all about whittling down diversity. New species emerged only once, back when organs were first combined in myriad ways. Many species subsequently went extinct, whereas the surviving species and their advantageous traits were preserved through the eons. Hippocrates, on the other hand, realized that evolution is a *creative* process that constantly produces diverse new species. To illustrate how new species can arise, Hippocrates described a peculiar custom of a race he dubbed the *Longheads* in his treatise *Airs, Waters, and Places*. According to Hippocrates, the Longheads considered elongated heads a status symbol. Consequently, Longheads used to bind the heads of babies in order to distort skull growth, resulting in children with abnormally long heads. Nature, Hippocrates went on to explain, reinforced this custom, and eventually babies were born with long heads, rendering artificial elongation unnecessary.[3] In other words, Hippocrates posited that organisms can inherit traits acquired by their parents during their parents' lives. This theory is now known as *Lamarckism* due to its erroneous attribution to Lamarck, a French zoologist who simply endorsed it over two millennia after Hippocrates' death.

Jean-Baptiste Pierre Antoine de Monet, chevalier de Lamarck, was born in 1744 to an aristocratic family of modest means in northern France. As the youngest of eleven children, Lamarck was originally destined for the priesthood. He attended a Jesuit college, but after his father's death and the 1761 expulsion of the Jesuits from France, Lamarck jumped at the opportunity to join the army. Unfortunately, Lamarck's promising military career ended abruptly when one of his brothers in arms playfully lifted him by the head and accidentally injured his neck. Lamarck subsequently dabbled in banking, music, and medicine before settling on a career in science. This decision was not as surprising as it may seem, because in Lamarck's day, science was all the rage in France. Back then, scientific books were more popular than books by Voltaire and Rousseau, and natural history cabinets replaced antiquities as the ultimate must-haves. Science was so trendy, in fact, that rich Parisians vied for electrostatic generators and kept hundreds of animal-shell merchants in business.* For Lamarck, who had first developed an ardent interest in botany as a soldier stationed in the south of France, choosing to devote his life to science may well have been a no-brainer.

Most modern scientists specialize in narrowly defined subfields, and often can't make head nor tail of scientific articles on topics that deviate from their area of expertise. The scientific

* Alfred Russel Wallace (1823–1913), who came up with a theory of evolution that was remarkably similar to Darwin's, had an agent who planned his trips to the tropics in accordance with the demand for specific specimens, and later sold Wallace's specimens to various collectors as well as the British Museum.

realm that Lamarck inhabited, however, revered multidisciplinarity, and Lamarck himself was at once a keen botanist, zoologist, paleontologist, physiologist, chemist, geologist, meteorologist, and physicist. Much like the philosophers of Ancient Greece, Lamarck sought to study nature in all its splendor in order to uncover the basic unifying principles underlying the universe. Consequently, whenever Lamarck pursued rigorous specialization, he did so with an eye to elucidating much broader truths.

Ironically enough, Lamarck was well respected by his contemporaries for the very accomplishments that he considered a means to a greater, cosmic end. His prolific classification of plant and animal species paved the way for his election to the French Académie des Sciences, and he consistently held prestigious scientific posts throughout the turmoil of the French Revolution and the Napoleonic era. One of these posts, professor of invertebrate zoology at the Museum d'Histoire Naturelle, was considered undesirable by Lamarck's colleagues. But Lamarck believed that studying the simplest of animals could shed light on the essence of life, and his research led him to formulate a new science, which he dubbed *biology*. This was nothing short of a major scientific breakthrough, because while his contemporaries tended to view botany and zoology as disparate domains, biology – in Lamarck's eyes – encompassed all living things, and was therefore uniquely suited to answering questions about our origins.

Lamarck's meticulous study of invertebrates led him to posit that inanimate matter can, under favorable conditions, give rise to the simplest of creatures. Life, according to him, is a force that binds the particles in our bodies to each other. When creatures die, their particles revert to inanimate matter, which can subsequently produce new life. But what happens to the simple creatures that arise from inanimate matter? Lamarck believed that "lower" creatures are constantly evolving to achieve higher levels of complexity, resulting in the breathtaking diversity of life forms populating the planet. Never one to shy away from controversy, he boldly asserted that even humans may have evolved from inferior animals. And if all creatures exist on an evolutionary continuum from inanimate matter to humanity, then the concept of species is no more than a construct deployed by scientists to make sense of reality.

The elegance of this hypothesis may have been derived from Lamarck's Deistic persuasion. Back in the seventeenth and eighteenth centuries, many prominent Frenchmen, such as Robespierre, Voltaire, and Rousseau, subscribed to *Deism*, the Enlightenment's most fashionable religion. Deism, which Robespierre had at one point managed to turn into the official religion of the French state, was a very peculiar belief system. Its adherents rejected all scriptures, miracles, revelations, and prophecies in favor of a transcendent God who was loathe to interfere in post-Creation earthly affairs. But why did the Deistic God take a step back after creating the world? According to the Deists, God had created the laws of nature, which, as scientists were discovering, govern the world perfectly. This, of course, renders the option of miraculous interventions unnecessary at best, and an affront to God's omnipotence at worst.

In Lamarck's Deist eyes, a rational creator had crafted a perfectly ordered evolutionary process, leaving nothing to chance. Once creatures were generated from inanimate particles, they had nowhere to go but up the ladder of complexity. But what kind/s of complexity would each creature pursue? Why do some creatures develop complex brains while others develop complex biochemical mechanisms to harness the sun's energy? Lamarck's answer to the latter all-important question is at the root of his status as one of the most misunderstood scientists in history. Like Hippocrates, he held that habits could influence heritable traits. According to Lamarck, the repetitive use of an organ causes what he called a "subtle fluid" to flow towards it, resulting in its reinforcement and incremental development. Differences in the distribution patterns of this subtle fluid affect the fluid that contains sperm, which explains how sperm comes to bear acquired traits. Conversely, when an organ is rarely used, the subtle fluid flows away from it, causing it to deteriorate and ensuring that the creature's sperm will transmit an atrophied organ to offspring.

These days, giraffe necks are the go-to example employed to explain Lamarckism, but Lamarck himself first used bird legs and feet to expound his ideas. Why, he asked, do swimming birds have webbed feet? The ancestor of swimming birds must have repeatedly spread the digits of its feet when attempting to move in the water, thus stretching the skin between its digits. This stretching was reinforced by the flow of Lamarck's subtle fluid, and webbed feet gradually developed. Similarly, the long legs of wading birds were acquired when an ancestral bird repeatedly stretched its legs to avoid sinking into mud. Over time, this habit elongated the bird's legs, producing the stilt-like limbs seen in modern wading birds. Lamarck also explained that the giraffe acquired its long neck by straining repeatedly to reach the leaves on trees, though he did not give this particular example much prominence in his writings. To demonstrate how disuse could lead to the disappearance of organs, he pointed out the vestigial eyes of subterranean moles and the surprising absence of legs on snakes, despite their close relation to legged animals.

Lamarck realized that evolution entails much more than organ growth or degeneration. Organs, he posited, could move around, as in the case of the flatfish's eyes moving to one side, or develop from scratch, as in the case of ruminants' horns and antlers. He also speculated that the arboreal ancestors of humans may have developed the ability to walk upright after venturing onto the ground, thus completing their ascension to the highest possible form of life. Lamarckism, then, was originally a comprehensive explanation of how and why life arose, gradually diversified into countless forms, and reached its zenith in humanity. Sadly, Lamarckism has since been reduced to the concept of heritable traits acquired through use – a concept that predated Lamarck and was universally accepted during his lifetime.[4] What we now call Lamarckism was for him little more than a self-evident detail incorporated into the epic story of life. But after his death, this almost insignificant detail came to overshadow his monumental contributions to science. As for the infamous giraffe example, it was repeatedly used to explain Lamarck's hypothesis, endlessly reinforcing the association of his name with a meager caricature of his ideas.

As we've seen, the ancients were avid evolutionists. Why, then, is evolution synonymous with Darwinism? What could Charles Robert Darwin, a nineteenth-century English naturalist, possibly add to the brilliant insights of Anaximander, Empedocles, and Hippocrates? And why has Lamarck's brand of evolution been cast aside? The answer is that Darwin's theory of evolution, and its fusion with the modern discipline of genetics, is so comprehensive that, as the esteemed biologist Theodosius Dobzhansky put it, nothing in biology makes sense except in the light of evolution.[5]

Let us start at the beginning. Darwin was born in 1809, the same year Lamarck published his theory of evolution. Fifty years later, Darwin, who had traveled around the world and collected specimens of over 1,500 species, published his revolutionary book *On the Origin of Species*. Life, the book argued, is a never-ending struggle, with organisms forced to compete constantly for finite resources. What differentiates winners from losers in this ongoing battle? According to Darwin, an organism's heritable traits can help or hinder survival. Birds with stronger beaks, for example, can crack more nuts, and will consequently tend to consume more nutrients. Similarly, zebras with longer legs will run faster when chased by lions, and will consequently tend to survive these chases. But what about weak-beaked birds and short-legged zebras? These unfortunate organisms are, of course, less likely to survive long enough to reproduce. And since organisms inherit traits from their parents, they will only rarely transmit their disadvantageous traits to offspring. The strong-beaked birds and long-legged zebras of the world, however, are much more likely to reproduce, and their advantageous traits are passed down from generation to generation.

As a country gentleman, Darwin was intimately familiar with selective breeding. This agricultural practice entails selecting plants or animals with desirable traits and breeding them together to reinforce these traits in the next generation. When a farmer selects the individuals that are bred, a process of *artificial selection* can be said to be taking place. But nature itself, Darwin realized, is a cosmic farmer tirelessly selecting individuals via the process of *natural selection.* Unlike Lamarck, Darwin did not claim that life seeks to increase its complexity. Natural selection, he believed, merely sorts through a variety of traits and favors the ones most conducive to survival amidst environmental adversity. Darwin's sublimely simple theory also held that life only had to arise once before branching out into innumerable forms. Small variations between these forms would thereafter be favored or penalized by natural selection, leading to survival or extinction, respectively.

Darwin's grandfather, Erasmus Darwin, had discussed the inheritance of acquired traits in his 1794 book *Zoonomia,** which may have inspired Lamarck to adopt the idea.[6] And although Darwin had originally championed natural selection as the driving force in evolution, he did include the heritable effects of use and disuse in later editions of *On the Origin of Species.*** So why do modern biology textbooks always contrast (factual) Darwinism with (erroneous) Lamarckism? The answer

* As Darwin recounted in his autobiography, he had read his grandfather's book twice and was disappointed to find that it was more speculative than fact based.

** In his autobiography, Darwin also discussed the effects of use and disuse on personal tastes. As a young man, Darwin had enjoyed poetry and other forms of art, but in later years he could not "endure to read a line of poetry". This, he lamented, seemed to be the result of his mind becoming "a kind of machine for grinding general laws out of large collections of facts". However, he believed that had he incorporated the reading of poetry into his weekly routine, the "atrophied" part of his brain responsible for "higher tastes" could have remained "active through use".[7]

to this question lies in the gradual evolution of the terms *Darwinism* and *Lamarckism*.

Lamarckism, as we've seen, has been reduced to a drawing of a primordial giraffe stretching its neck to reach the leaves of a tree. It also hasn't helped that Lamarck's name was unfairly linked to the policies of Trofim Lysenko, a Stalin-era agronomist who sought to transform Soviet agriculture with his "Marxist genetics". Lysenko, you see, believed that acquired traits are heritable, though he vehemently denied any association with Lamarckism. When he became the director of the Institute of Genetics of the Soviet Academy of Sciences, he turned his many pseudoscientific ideas into state-sanctioned doctrine and forcibly suppressed criticism. Hundreds of scientists who disagreed with him were imprisoned, and several were even executed. Lysenko then forced Soviet farmers to adopt his agricultural methods, which led to severe famines that killed millions of people. Undeterred by these disastrous results, Mao Zedong later adopted Lysenkoism, a decision that led to a further fifty-five million people losing their lives in what is now known as the Great Chinese Famine of 1959–1961.

While Lamarck's name was being dragged through the mud, Darwin's theory was purged of any hint of acquired inheritance and fused with the rapidly developing field of genetics, yielding *neo-Darwinism*. But what, exactly, is neo-Darwinism? To answer this question, let us examine what Darwin knew, as well as what he couldn't possibly know. Darwin knew that in any given species, individuals will always possess slightly different traits. But he didn't know *why* this variation existed in the first place. Similarly, he knew that many of the traits that differentiate individuals are heritable, but he had no idea *how* these traits were passed down to offspring. This is where modern genetics comes in. We now know that each individual's basic traits are encoded in a chain-like molecule called *DNA*. Individuals differ because differences in their DNA produce different traits, and when they reproduce, they replicate their unique DNA and transmit it to their offspring, thus allowing them to pass down parental traits. We also know that *mutations*, the changes that are constantly occurring in DNA, can alter traits. This last point is key to our understanding of evolution, because it means that natural selection is primarily concerned with the sorting of advantageous and disadvantageous mutations.

Genetics, as you can see, was Darwinism's missing puzzle piece. With its incorporation, neo-Darwinism ascended to its lofty perch in the scientific consensus, where it remains to this day. Interestingly, the notion of random mutations quickly became a canonical component of neo-Darwinism. Unlike Lamarckian organisms, which constantly strove for increased complexity, neo-Darwinian organisms, it was decided, must mutate at random, rendering evolution a natural process free of any guiding forces. But are mutations truly random? Darwin himself expostulated that chance alone could explain the variation he observed in organisms. "I have hitherto sometimes spoken as if the variations – so common and multiform in organic beings under domestication, and in a lesser degree in those in a state of nature – had been due to chance," he wrote in *On the Origin of Species*. "This, of course, is a wholly incorrect expression, but it serves to acknowledge plainly our ignorance of the cause of each particular variation."[8]

CHAPTER 2

Mutation Menagerie

Mutations, as we've seen, are changes that occur in DNA. But why do these changes have the power to alter our traits? To answer this question, we must first take a deep dive into a subcellular world inhabited by three wondrous molecular chains. The first of these is DNA, a chain composed of four kinds of molecules called *nucleotides*. The DNA nucleotides, adenine (A), cytosine (C), guanine (G), and thymine (T), are strung together in seemingly endless chains that are bundled up for safekeeping in cells. Importantly, each DNA chain is linked to another DNA chain all along its length, forming *double-stranded* DNA. In this unique structure, an A in one strand is always linked to a T in the other strand, and a C in one strand is always linked to a G in the other strand. Consequently, the two strands possess *complementary* sequences. Before a cell divides, it replicates its DNA by "unzipping" the two strands and building a new, complementary strand for each, thus generating two copies of the cell's hereditary information that can be divvied up when the cell splits in two.

But why do cells need DNA in the first place? This brings us to our next molecular chain – RNA. Like DNA, RNA is a chain of nucleotides, although its nucleotides differ from the ones that make up DNA. RNA is also usually *single-stranded*, for reasons that will soon become clear. Every so often, a small segment of double-stranded DNA is unzipped, exposing a short sequence known as a *gene*. A complementary RNA sequence is then built on the gene via a process called *transcription*. After transcription, the short RNA sequence breaks free and the DNA is zipped up once again. The RNA then undergoes *translation*, which brings us to our final molecular chain – protein. Proteins, you see, are long chains of molecules known as *amino acids*. To build proteins, every RNA nucleotide triplet, or *codon*, is translated into a specific amino acid that is added to a growing chain. The resulting proteins are key to all of the molecular structures and mechanisms maintaining life. As enzymes, they catalyze vital chemical reactions. As antibodies, they fight off infections. They also enable muscle contraction, make up many biological structures, and facilitate intercellular communication. Proteins, in short, determine an organism's traits, and since their sequences are ultimately derived from DNA sequences, mutations in DNA can drastically alter traits.

Having established the long reach of mutations, we turn to a pressing question: why do mutations occur in the first place? There are multiple answers to this question. Some nucleotides can spontaneously shapeshift for a fraction of a second, temporarily altering the DNA sequence. This may not seem to matter in the grand scheme of things, but if these "quantum jitters" occur just as a nucleotide is being replicated, the newly created strand of DNA will accidentally incorporate a wrong nucleotide into its complementary position, yielding a permanent mutation.[1]

DNA decomposition is also a source of mutation. Like all large biological molecules, DNA can spontaneously decompose for a number of reasons. Exposure to oxygen, for instance, can cause *oxidation*, which leads to a great deal of DNA damage. This is evident in rats, where oxidation is responsible for around two thirds of all mutations.[2] So severe is the threat of oxidation-induced mutations, in fact, that many cells safeguard their DNA in a poorly oxygenated compartment known as the *nucleus*. Interestingly, nucleated cells also restrict many chemical reactions that may lead to oxidation to their *mitochondria*, little organelles that possess their own DNA. What, you may be wondering, happens to this *mitochondrial DNA* as a result of excessive oxidation? The answer is that it decomposes rapidly and is consequently characterized by a very high mutation rate.[3]

Mutations can also be caused by chemical or physical agents known as *mutagens*. This category includes cigarette smoke, UV radiation, and various chemicals. It also includes viruses and other similar entities, which will be discussed at length in Chapter 8. But even if we could somehow remove all mutagens from our environment, we would still be plagued by mutations because many mutations originate from copying errors. Remember that when cells prepare for division, they replicate their DNA. This way, every cell division produces two cells, each with its own copy of the parent cell's DNA. Ideally, the two copies would be identical, but replicating long sequences perfectly is virtually impossible, and errors will inevitably occur in every replication cycle.

To understand why copying errors are a fixture of DNA replication, imagine a guslar reciting a long poem. Would we expect the guslar to recite the poem exactly the same way every time? Not really. As we've seen, the guslari were prone to tweaking their poems with every recitation. Consequently, one letter could easily be swapped for another, as in the case of *cat* → *bat*. Similarly, during DNA replication, one nucleotide is occasionally replaced by another in what is known as a *substitution mutation*. The reason? DNA replication is carried out by a family of fallible enzymes called *DNA polymerases*. These enzymes build new strands of DNA as they glide along existing strands and "read" their sequences. Most DNA polymerases take pains to incorporate a complementary nucleotide into every position of a growing strand, but (infrequent) mistakes are unavoidable. When the bacterial DNA polymerase I, for example, builds a new strand of DNA, it will add a wrong nucleotide to a sequence every 10,000 to 100,000 nucleotides. Curiously, its accuracy seems to be influenced by nucleotide size, because when researchers synthesized artificial nucleotides of various sizes, DNA polymerase I was a whopping twenty-two times more efficient at incorporating artificial nucleotides that were slightly larger than natural nucleotides.[4]

This finding is significant because it suggests that there is plenty of wiggle room for the wrong nucleotide to slip through.

And speaking of wiggle room, there also seems to be wobble room, for lack of a better term. Francis Crick, who was awarded a Nobel Prize for his role in deciphering the double helical structure of DNA, hypothesized that many substitution mutations occur due to an almost imperceptible shift in a nucleotide's position in three-dimensional space. This *wobble*, as it is known, allows the nucleotide to bind a non-complementary nucleotide in the opposite strand, yielding a mutation.[5]

Interestingly, the effect of substitution mutations varies due to the intricacies of the *genetic code*. Amino acids, as we've seen, are encoded by nucleotide triplets, or *codons*. But since four types of nucleotides can be arranged in sixty-four different codons, it is somewhat surprising that proteins are composed of just *twenty* types of amino acids. How, then, are the "excess" codons utilized? Many of them are synonymous, which means that the same amino acid can be encoded by multiple codons. If a mutation results in the substitution of a codon with a synonymous codon, it is a *silent mutation*. But if the mutation does manage to replace one amino acid with another amino acid, it is a *missense mutation*. There are also three codons that don't encode amino acids at all. These codons are known as *stop codons* because they mark the end of the protein encoding sequence. Naturally, a substitution mutation that replaces an amino acid encoding codon with a stop codon will produce a truncated protein, which is why these mutations are called *nonsense mutations*.

At first glance, nonsense mutations appear to be highly deleterious. After all, many of them are *loss-of function mutations*, or mutations that reduce or eliminate a gene's activity entirely. But a nonsense mutation may simply inactivate a gene until a future mutation reverses the nonsense mutation and revives it. Bacteria, for example, often possess inactive genes that encode enzymes required for the utilization of various sugars. When researchers grow such bacteria in a medium containing a sugar they can't utilize, a mutation quickly reverses a loss-of-function mutation,[6] restoring an inactive gene to its former glory. Similarly, silent mutations, which appear to be benign, can actually be remarkably influential, because although they do not alter the protein produced by a gene, they sometimes change the number of copies of the protein that gene can produce. How is this possible? One eye-opening study on the subject demonstrated that silent mutations can transform the way transcribed RNA folds in three-dimensional space. And if a silent mutation causes RNA to form a tightly folded structure that is inaccessible to the cell's translation machinery, fewer proteins will be churned out.[7]

To add insult to injury, a silent mutation in one gene can also affect the expression of other genes. In one revealing study, researchers designed nine synonymous versions of a certain gene and introduced them into identical bacteria. Each version of the gene was supposed to yield the same outcome, but in practice, some versions increased the gene's protein production – as well as that of its neighboring genes. Changing the DNA sequence, as it turned out, sometimes creates new binding sites for *RNA polymerase*, the enzyme that transcribes DNA into RNA. And when RNA polymerase latches on to these unorthodox binding sites, it cranks up transcription of both the mutated gene and additional genes in its vicinity.[8]

Let us return to the guslar reciting a poem. How else would we expect him to change the poem? Guslari did not memorize poems letter by letter or even word by word; they were much more likely to memorize a line at a time. But if a line happened to be particularly boring, they would, over time, omit it, or at least part of it. Alternatively, they could add a line or two to spice things up. Similarly, *insertion mutations* and *deletion mutations* expand and shorten DNA sequences, respectively. This happens when a strand of DNA slips as it is being elongated, causing DNA polymerase to add extra nucleotides to it, or when the copied strand slips as it is read, causing DNA polymerase to omit nucleotides.

Slippage, incidentally, is very common in genomic regions that contain many short, repetitive sequences, where DNA polymerase can easily lose its place. And when this happens, the DNA sequence gains or loses a nucleotide, a codon, or even many codons. But not all insertion or deletion mutations are created equal. Since codons are *triplets* of nucleotides, the insertion or deletion of one nucleotide will shift all following codons, rendering the rest of the amino acid sequence unrecognizable. This is called a *frameshift mutation*, and it occurs whenever the number of inserted or deleted nucleotides is not divisible by three. However, if the number of inserted or deleted nucleotides *is* divisible by three, the resulting protein will contain extra amino acids or lack amino acids but will otherwise remain unchanged.

Every once in a while, a guslar would make a major change to a poem, such as eliminating a stanza or adding in a brand-new one. Major changes, or errors, of this sort are also known to occur in DNA replication. Genomes, you see, are typically composed of multiple DNA molecules known as *chromosomes*, which are replicated simultaneously before cell division. The sheer complexity of this massive operation occasionally results in *chromosomal mutations*, which include the fusion of chromosomes, the exchange of segments between chromosomes, and the deletion, duplication, or inversion of segments in a single chromosome. Inversion mutations, incidentally, possess the mystifying potential to drastically alter body size, as demonstrated by Brenden Adams, who at seven feet and eight inches once held the Guinness World Record for tallest teenager. Brenden, as it turned out, did not achieve his formidable stature by eating all his veggies and getting plenty of sleep – he simply had an inversion mutation on chromosome 12!

Chromosomal mutations are, in many ways, the greatest drivers of evolution. This is not to say that smaller mutations cannot make their mark. Cystic fibrosis, for example, is a severe genetic disorder caused by one small mutation. Just how small can this life-changing mutation be? Well, of the over 2,000 cystic fibrosis mutations described to date,[9] many are simple missense, nonsense, or frameshift mutations.[10] Still, it is the chromosomal mutations that have the greatest evolutionary potential. Take, for instance, the chromosomal mutation that determines the number of chromosomes in each of your cells. The human genome consists of twenty-three pairs of chromosomes, though other Great Apes have twenty-four pairs. Does this mean that our closest cousins acquired an extra pair of chromosomes? Or did we somehow lose a pair? The solution to this mystery lies in the peculiar structure of human chromosome 2. Part of this chromosome's sequence is almost identical to the sequence of one chimpanzee chromosome, and another part is almost identical to the sequence of a different chimpanzee chromosome, suggesting a chromosomal fusion of some sort. Moreover, although most chromosomes possess *telomeres*, or regions of repetitive nucleotide sequences, at their ends, human chromosome 2 has two additional (vestigial) telomeres right in its middle. It is therefore clear that an ancient chromosomal mutation resulted in the end-to-end fusion of two chromosomes.[11] This mutation is estimated to have occurred almost one million years ago,[12] after our ancestors diverged from the ancestors of chimpanzees, but before our divergence from the Neanderthals and Denisovans, who shared our fused chromosome.[13]

Chromosomal mutations are also vital sources of evolutionary innovation due to their ability to duplicate long stretches of DNA. Whenever this happens, entire genes are duplicated, whisking redundant copies of genes into existence. This redundancy is actually liberating, because the evolution of novel functions is often limited by the need to maintain existing functions. Put simply, if a vital gene acquires mutations that impair its function, the organism that possesses this gene will die, even if these mutations could potentially lead to the development of a novel adaptive function. But when a gene is duplicated, one copy stays behind to maintain its "traditional" function, while the other copy is free to evolve in new directions. Consequently, duplicate genes mutate much faster than unduplicated ones, a process that sometimes leads to *neofunctionalization*, or the creation of a new function for a gene.

Good examples of neofunctionalization can be found in *globins*, a family of genes generated by a series of gene duplications. Human readers are probably most familiar with the globin genes encoding *hemoglobin*, the protein complex that shuttles oxygen around our bodies. But these genes are just part of the story. The hundreds of globin genes found in everything from bacteria to plants and animals[14] have neofunctionalized repeatedly throughout evolutionary history, mutating to fill roles as diverse as oxygen storage and light detection.[15]

The most fascinating globin gene duplication is undoubtedly the one that gave rise to the hemoglobin genes, but to understand what makes this gene duplication so extraordinary, we must first familiarize ourselves with one of hemoglobin's lesser-known cousins – myoglobin. The myoglobin gene encodes an eponymous protein with a single oxygen-binding site. This protein's main function is supplying muscle cells with oxygen, which it does by binding oxygen tightly and storing it until strenuous muscle activity requires its release. Prior to the evolution of the circulatory system, an oxygen-storing protein was sufficient for survival, which is why the earliest animals possessed a myoglobin-like gene. But hundreds of millions of years ago, this gene was duplicated in the course of whole-genome duplication,[16] and one of the resulting copies neofunctionalized to become the ancestor of all hemoglobins. Unlike myoglobin, hemoglobin is not a single protein, but a protein complex consisting of four proteins – a *tetramer*. Each of these proteins has an oxygen-binding site, which means that hemoglobin can bind four times as much oxygen as myoglobin. Hemoglobin also differs from myoglobin in its affinity for oxygen, which is much lower, meaning that hemoglobin can't bind oxygen as tightly as myoglobin does.[17]

At first glance, this inferior affinity to oxygen appears to be disadvantageous, but is it really? In almost all vertebrates, hemoglobin binds oxygen in the respiratory system and then disperses it to the rest of the body as it cruises through capillaries. The ability to release oxygen easily, then, is an obligatory requirement for hemoglobin, which enabled early vertebrates to evolve a circulatory system. The circulatory system, in turn, enabled vertebrates to evolve massive bodies, because it allowed internal organs, which have no access to atmospheric oxygen, to receive plenty of oxygen from the bloodstream.

Intriguingly, when researchers reconstructed the sequence of the primordial myoglobin-like gene discussed earlier, they discovered that it would have had to undergo very few mutations to produce hemoglobin. Following gene duplication, just two substitution mutations would have sufficed to cause the formerly solitary protein to *tetramerize*, or form a four-protein complex. These same two substitution mutations, incidentally, also drastically reduced oxygen affinity,[18] paving the way for the development of hemoglobin's unique role in vertebrate biology.

Neofunctionalization is not the only possible outcome of gene duplication. Following gene duplication, a gene may acquire deleterious mutations, but only if the other copy of that gene complements its newfound defects. Thus, the ancestral gene's function may come to be distributed among its copies via the process of *subfunctionalization*. For good examples of subfunctionalization, look no further than the human hemoglobin genes, which we will now examine more closely.

Hemoglobin, as we've seen, is a protein complex composed of four proteins, the identity of which we previously glossed over. Two of these proteins are identical *α-globins*, and the other two are identical *β-globins*. Curiously enough, neither α-globins nor β-globins can form hemoglobin on their own; two of each are required for tetramerization. But this wasn't always the case; primordial β-globins used to tetramerize all by themselves. That all changed when the β-globin gene underwent duplication, yielding what would become the α-globin gene. Both the β-globin gene and the α-globin gene subsequently underwent a process of subfunctionalization, resulting in their obligatory interdependence.[19]

But that's not all – at some point in our distant evolutionary past, gene duplications distributed hemoglobin's oxygen-shuttling function among various gene copies that came to be expressed at different developmental stages. That's how we ended up with embryonic hemoglobin, which consists of either two ζ-globins and two ε-globins or two ε-globins and two α-globins; fetal hemoglobin, which consists of two α-globins and two γ-globins; and adult hemoglobin, about 98% of which consists of two α-globins and two β-globins, and about 2% of which consists of two α-globins and two δ-globins. This elaborate subfunctionalization ensures that hemoglobin meets the specific needs of each developmental stage. Fetuses, for example, develop within oxygen-poor uteruses, and must therefore obtain as much oxygen as possible from the maternal bloodstream. Adults, on the other hand, can inhale a virtually endless supply of oxygen, and can stand to "forfeit" some of it. Consequently, fetal hemoglobin is characterized by a much higher oxygen affinity than adult hemoglobin,[20] which, as we've seen, cannot afford to bind oxygen too tightly. Naturally, a newborn must complete the fetal-hemoglobin-to-adult-hemoglobin transition quickly to survive in its new environment, which is why newborns break down fetal hemoglobin en masse. But if a newborn's liver struggles to clear the products of this massive breakdown, they will accumulate in the blood,[21] resulting in the yellowing of the skin and eyes – an extremely common condition known as *neonatal jaundice*.

Adapting to life in the extrauterine world is not the end of the road for adult hemoglobin, because the extrauterine world is a highly dynamic environment to which genes must constantly adapt. Malaria-causing *Plasmodium* parasites, for instance, have evolved to feed on hemoglobin, necessitating further evolution on the hemoglobin genes' part. But how, exactly, can hemoglobin mutate to avoid becoming a parasite's snack? To answer this question, let us examine how *Plasmodium* accesses hemoglobin in the first place. Hemoglobin, it should be noted, does not course through our blood vessels on its own. Instead, it fills our red blood cells, which squeeze through even the thinnest of capillaries in their quest to deliver oxygen to all tissues. Now, when a *Plasmodium*-infected mosquito feeds on a human's blood, it unwittingly injects parasites into its victim's bloodstream, and these parasites waste no time in invading red blood cells and gorging themselves on hemoglobin. In sub-Saharan Africa, where malaria is endemic, dying of malaria is an acute annual threat.[22] Fortunately, many people of sub-Saharan descent possess a β-globin gene with a missense mutation that yields a deformed β-globin. This deformed protein, in turn, causes some of the red blood cells it inhabits to collapse into a sickle-like shape. And if *Plasmodium* happens to invade one of these sickle cells, it is killed instantly by a dearth of oxygen. Thus, the sickle cell mutation cunningly protects roughly 90% of its carriers from severe malaria.[23]

Speaking of mutations that confer malaria resistance, the membranes of most human red blood cells contain *Duffy receptors*, which play an important role in the immune system's inflammatory response. Duffy receptors also happen to be crucial to *Plasmodium* invasion because each *Plasmodium* parasite must latch on to a Duffy receptor before entering a cell. It is therefore unsurprising that over two thirds of people of African descent lack Duffy receptors on their red blood cells, unlike the vast majority of individuals from other ethnic groups.[24] How did this malaria resistance trick arise? As it turns out, a single nucleotide substitution mutation was all it took to abolish the Duffy receptors, leaving *Plasmodium* unable to penetrate red blood cells.[25]

CHAPTER 3

Forgettable Flukes

The word "mutation" usually conjures up all of its most terrifying consequences, from cancer to ninja turtles. But the vast majority of mutations are actually unlikely to make much of a difference, as demonstrated by the 200 or so mostly inconsequential mutations every generation adds to the human genome.[1] Amazingly, these mutations are just the tip of the iceberg; the vast majority of mutations are swiftly corrected and leave behind no trace of their transient existence. Most of the mutations that pop up during DNA replication, you see, are corrected in real time via the aptly named process of *proofreading*, which replaces incorrectly incorporated nucleotides in the growing strand.

Given that roughly 99% of mutations are abolished in this manner, one would be forgiven for assuming that the rare mutation that somehow escapes proofreading is off the hook. But nothing could be further from the truth. After replication, the *DNA mismatch repair* system kicks into gear. This complex system, which employs multiple proteins, identifies areas in the double-stranded DNA where mutations have led to incorrect nucleotide pairings, thus distorting the double helix. Once a mutation is identified, the system will snip out an erroneously incorporated nucleotide – as well as up to thousands of adjacent nucleotides – before patching up the gap with the right nucleotides. Although mismatch repair is highly efficient, a few mutations do manage to slip through the cracks, and it is these mutations that end up changing the genome permanently.

What exactly do these persistent mutations do? Generally speaking, not much. Most mutations cannot affect us simply because they are not in *coding sequences*, or sequences that contain genes. Only about 1% of the human genome consists of coding sequences, which means that the chances of a mutation occurring in a gene are vanishingly small. And even if a mutation *does* occur in a gene, it won't necessarily make much of a difference because, as we've seen, some mutations are silent mutations, or mutations that replace a codon with a synonymous codon, leaving the amino acid sequence unaltered.

But what if the mutation does change the amino acid sequence? Surprisingly, even missense mutations are often insignificant. Proteins, you see, are not just chains of amino acids, they're folded chains of amino acids – each protein has a unique three-dimensional fold upon which its functionality depends. Some segments of the amino acid chain will fold in a certain way to serve a certain purpose, such as binding other molecules, cleaving molecules, or joining molecules. These functionally invaluable areas in the protein's structure are known as protein *domains*. The areas between protein domains, unsurprisingly, aren't nearly as significant as the domains, meaning that if a missense mutation occurs between domains, it is unlikely to affect protein function. Indeed, it seems that changing the amino acids in most positions along a protein's sequence will have no impact whatsoever, as demonstrated by a strikingly simple experiment carried out on a minuscule virus.

The *Escherichia virus T4* infects *Escherichia coli*, or *E. coli*, bacteria. In the course of a successful infection, T4 multiplies within *E. coli* before using a viral protein known as a *lysozyme* to break down the bacterial cell wall, thus killing the bacterium to enable the release of a swarm of new viruses. Consequently, when T4 is introduced to a dense layer of *E. coli* on a petri dish, little holes, or *plaques*, will appear wherever a successful infection occurs. Plaque formation depends on the existence of a functional lysozyme protein, because without it, T4 is incapable of killing *E. coli*. So when researchers created a massive collection of T4 lysozyme genes, each with a missense mutation at a different position, testing the various mutant genes for plaque formation was the obvious way to determine lysozyme functionality. Incredibly, well over half of the positions in the lysozyme's sequence tolerated amino acid substitutions,[2] proving that genes can get away with a large number of missense mutations.

But what if a missense mutation occurs in a crucial position, such as one responsible for interacting with another protein? This, too, may not be cause for concern. The twenty amino acids that make up proteins can be divided into just four groups of amino acids with similar properties, meaning that if a missense mutation replaces one amino acid with another amino acid from the same group, their similarity will ensure a negligible effect on the protein's functionality.

This leaves us with missense mutations that replace amino acids in key positions with radically different amino acids. Such mutations must surely have far-reaching implications! However, a closer examination of how genes work in real, living organisms sheds light on why even this is not necessarily true. As we've seen in the case of the various human hemoglobin genes, genes are not constantly churning out proteins – most genes are only expressed in a clearly delineated developmental stage. Many human genes, for instance, encode proteins with roles in embryonic development which cease to be expressed once embryogenesis is completed. Consequently, if a mutation happens to drastically alter one of these genes later in life, it will go unnoticed *despite* its immense destructive potential. Moreover, most of our genes are not expressed in every single cell in our bodies. Lots of genes are tissue-specific, which means that different tissues express different subsets of genes. So if a mutation in a cell in one tissue alters a gene that is only expressed in a different tissue, this, too, will prove inconsequential.

Nonsense mutations and deletion mutations are trickier, because it is difficult to imagine how a truncated protein can be anything but disastrous. But what if the nonsense mutation or deletion mutation occurs, say, just a few amino acids before the protein's end? Under these circumstances, the mutation is unlikely to shorten or eliminate any important domains, allowing it to slip under the radar. And if it *does* occur in a more prominent position, it may still not be a big deal because even loss-of-function mutations don't necessarily matter in the grand scheme of things. Take, for instance, the O blood group. All humans have either type A blood, which means that they have "A" sugars on the surface of their red blood cells; type B blood, which means that they have "B" sugars on the surface of their red blood cells; type AB blood, which means that they have both; or type O blood, which means that they have neither. Type O blood is not a "real" type; it is simply the result of the deletion of the genes behind the A and B sugars.[3] This loss-of-function mutation, however, is wholly benign, as any readers with type O blood can attest.

Once in a blue moon, a mutation will clear all of the hurdles discussed earlier to actually affect the cell it inhabits. But even an extraordinary mutation of this sort will probably be unremarkable, evolutionarily speaking, unless it is passed down to the next generation. Humans (and other multicellular organisms) possess two kinds of cells: the precursors of sperm and egg cells, or *germ cells*, which produce offspring, and *somatic cells*, which make up the rest of our bodies. The larger an organism's body, the more numerous its somatic cells will be, which means that a greater percentage of its mutations will occur in somatic cells. And unlike germline mutations, which can be passed down to offspring, somatic mutations have no way out of the bodies in which they arise. In other words, somatic mutations are, by definition, incapable of making an evolutionary mark, regardless of how influential they may be within the confines of a body. Moreover, truly egregious somatic mutations are frequently stamped out before getting the chance to cause any trouble. The occurrence of cancerous mutations, for example, usually triggers *apoptosis*, a kind of cellular suicide that averts tumor development.[4]

Meanwhile, the germline mutations, which, as we've seen, are the only ones with the power to make a real difference, seem to be actively suppressed. One fascinating study, which compared the somatic mutation rate with the germline mutation rate in both humans and mice, revealed that the mutation rate in somatic cells is approximately two orders of magnitude higher than the mutation rate in germ cells.[5] This astonishing disparity implies that germ cells work to suppress the occurrence of mutations, perhaps by utilizing enhanced DNA repair mechanisms. And if germline mutations, the only mutations capable of generating evolutionary innovation, are largely prevented from doing so, it is a wonder that evolutionary changes occur at all!

As it turns out, even the accrual of mutations is often far from calamitous. Consider the widely held belief that the accumulation of mutations over many years underlies the cellular deterioration associated with aging.[6] At first glance, this hypothesis appears entirely reasonable. Organisms *do* accumulate mutations throughout their lives, and mutations clearly possess the power to impair all manner of cellular processes. But one groundbreaking study on the subject found that normal human cells routinely tolerate elevated mutation burdens without exhibiting any signs of premature aging.[7] This counterintuitive finding indicates that even multiple mutations somehow frequently fail to take a toll on cellular functionality.

Why can organisms afford to amass so many mutations? The answer to this puzzling question might have something to do with the existence of numerous apparently inessential genes which can mutate without repercussions. Take the human genome, with its 20,000 or so genes.[8] How many of these genes do we really need? Not that many, one could argue, especially since organisms with smaller genomes grow faster due to a more efficient replication process and a lower demand for the nutrients that constitute DNA building blocks.[9] And given that yeast does very well with around 6,000 genes,[10] one would be forgiven for assuming that the human genome is unnecessarily bloated, whereas the yeast genome contains the bare minimum of genes required for life. This, however, is patently false, because a staggering 85% of yeast genes can be disrupted by mutations with no effect on viability![11]

Organisms, then, are extremely tolerant of the very few mutations that somehow manage to alter their proteins. Paradoxically, it is the *more* mutation-tolerant genes that tend to produce significant evolutionary innovation, such as the higher-level cognition we humans are so proud of. This was demonstrated in a study that identified over a hundred genes that are expressed in all primate brains, albeit very differently in human brains and the brains of other primates. The human versions of these key genes, it turns out, have long been exceptionally tolerant of mutations. But rather than hindering evolution, their unique capacity to carry on as usual while being bombarded by mutations has enabled the acquisition of uniquely advanced cognitive abilities.[12]

CHAPTER 4

Predictable Patterns

Having established how insignificant most mutations are, let us now turn to their perceived fortuitousness. As we saw in Chapter 1, the notion of random mutations has come to be considered a core component of neo-Darwinism. But are mutations really random? Nature is rarely as dogmatic as those who study it. For starters, the word "random" is, in a way, almost devoid of scientific meaning, because every natural phenomenon is preceded by events that play a role in shaping it. Mutations, when examined through this lens, will never be independent of the molecular circumstances of their birth, and can therefore never be truly random.

This isn't just a theoretical argument; examples of nonrandom mutations abound in complex multicellular organisms and simple microbes alike. Take, for instance, the lowly *ΦX174*, a tiny virus with a genome of just eleven genes. In one illuminating experiment, researchers sequenced the genomes of two ΦX174 populations before propagating them separately for ten days – the equivalent of about one thousand generations. When the genomes of the two populations were once again sequenced at the end of the experiment, it emerged that roughly half of all of the mutations acquired throughout the experiment were identical in the two populations.[1] Now, if mutations were completely random, wouldn't the viruses in each population have ended up with a completely different set of mutations?

This kind of unnerving nonrandomness is also apparent in the mutations of another, better-known virus: HIV. When researchers sequenced a key HIV gene in the viruses infecting eight patients before the initiation of drug therapy and after the experiment's completion, they found that this gene had evolved to neutralize the drug's effects in five of the patients. And amazingly, the viruses infecting all five of these patients had acquired the same missense mutations at the same two positions along the protein's sequence![2] An additional form of nonrandomness was demonstrated in a different study focusing on a different HIV gene which frequently evolves to confer resistance to another drug. When eighteen HIV patients were treated with that drug, the viruses in most patients responded by acquiring the same four mutations in the investigated gene, *in the same order*.[3] In summary, the mutations that occur in the genomes of distinct viruses, as well as the order in which they occur, can be uncannily similar.

Discovering that small mutations in simple viral genomes are nonrandom is all well and good, but can the same be true for larger mutations in more complex genomes? Are large-scale chromosomal rearrangements, such as the deletion, duplication, or inversion of long stretches of DNA, actually random? This poses a critical question, because large-scale chromosomal rearrangements are exceedingly likely to have enormous evolutionary implications. They also occur extremely infrequently, which is why researchers who wished to determine their randomness (or lack thereof) propagated twelve identical populations of *E. coli* bacteria separately, but under the same conditions, for a whopping twenty-five years! After evolving for forty thousand or so generations, the genomes of the different populations were examined, and lo and behold, most large-scale chromosomal rearrangements were found to have occurred in at least two of the twelve populations.[4] Given the rarity of such colossal mutations, this discovery is a clear indicator of overt nonrandomness.

Nonrandom mutations can also result in confusing classifications, as in the case of the dysentery-causing *Shigella* bacteria. *Shigella*, you see, is the name of a bacterial genus, which would normally mean that all of its strains are descended from a common ancestor, and therefore possess very similar genomes. But in the peculiar case of the *Shigella* genus, the different strains actually arose independently and converged on similar genomes! We know this because sequencing key bacterial genes in forty-six *Shigella* strains revealed that massive mutations have led to the evolution of a *Shigella*-type genome over and over again.[5] This, of course, indicates that the *Shigella* genus is not a real genus, and that there is nothing random about the mutational path to "Shigellahood".

The nonrandom nature of the mutations discussed thus far has been deduced by comparing the similar evolutionary trajectories of distinct populations and asking a simple question: how likely is such similarity to be a coincidence? But genuine nonrandomness shouldn't be limited to inter-population comparison; it should also be apparent in the mutations of *individuals*. Put simply, for mutations to be deemed unequivocally nonrandom, an individual's mutations must be shown to be predictable.

But how can this be done? We can't rewind our lives to see if we acquire the same mutations again, can we? As humans, no. Bacterial lives, however, can be rewound, in a way, due to three important differences between bacteria and humans. The first is that each bacterium is composed of a single cell, as opposed to the approximately thirty-seven trillion cells that make up a human body.[6] This is important, because it means that unlike humans, who can acquire a mutation in just one cell and fail to pass it on to the next generation, the bacterial cell will always pass down its mutations. The second difference is that bacteria clone themselves to produce offspring, whereas human reproduction entails the mixing of about half of one human's genes with the genes of another human. This, too, is important, because it means that every bacterium is virtually identical to its "parent". Last but not least, bacteria can be frozen and revived easily. Humans, on the other hand, can be frozen, but we have yet to figure out how to revive them.

So what do these differences have to do with rewinding life? Well, since bacteria clone themselves every time they reproduce, a population produced by a single bacterium is essentially a collection of clones of the same individual. If one takes some of these clones and freezes them, and allows the rest to continue reproducing for, say, one hundred generations before reviving the frozen clones, the frozen clones will then relive the one hundred generations *already experienced* by their unfrozen counterparts. Put simply, every time scientists revive frozen bacteria, they are rewinding the lives of their closest kin.

This trick was employed by the researchers who propagated twelve populations of *E. coli* bacteria in a years-long experiment. Although *E. coli* primarily lives off glucose and cannot consume citrate under normal conditions, the twelve populations were grown in a glucose-limited medium with citrate, which gave them an evolutionary incentive to acquire citrate-utilizing mutations. The researchers froze a sample of bacteria from each population every 500 generations, and waited. Eventually, after about 31,000 generations, one of the populations started utilizing the citrate in its medium. Life was then rewound by reviving samples frozen at different stages of that population's history, and the citrate-utilizing trait re-evolved in many of the samples frozen after the 20,000th generation mark.

How can we make sense of this remarkable outcome? It seems that the ability to utilize citrate requires multiple mutational steps, the last of which depends on a mutation that occurred after 20,000 or so generations.[7] Thus, samples frozen after the appearance of this potentiating mutation readily acquired the final mutation necessary to consume citrate, proving that even complex evolutionary processes can be brazenly nonrandom.

Incredibly, millions of years before the conduction of this experiment, nature itself conducted a similar rewinding experiment, albeit with a very different kind of freezing. This experiment took place in Antarctica, which was once part of a lush, dinosaur-inhabited supercontinent near the equator. About eighty million years ago, Antarctica broke free and drifted southwards until it settled in its current location over the South Pole. It was there that Antarctica's lakes began to freeze in the winter, but up until roughly fifteen million years ago, the ice would still melt in the summer. Eventually, however, the ice stopped melting and gigantic glaciers glided over Antarctica, yielding approximately 400 subglacial lakes.

One of these lakes, now known as Lake Vostok, was discovered when Russian researchers drilled into a glacier and discovered a *liquid* lake trapped under about two and a half miles of ice. Geothermal energy had prevented Lake Vostok's water from freezing for millions of years, raising an intriguing question: could any of the lake's ancient inhabitants have survived in a pitch-black body of water cut off from the rest of the world and subjected to over 300 times its original pressure?

Any organisms still living in Lake Vostok would probably be unicellular, so rather than embarking on a subglacial fishing trip, researchers searched for DNA fragments in ice obtained from the lake's roof. Thus despite its seemingly inhospitable conditions, Lake Vostok was found to contain well over 3,000 unique DNA sequences, which means that it is inhabited by at least that many species. And interestingly, many of these species' gene sequences were shown to be identical to the gene sequences of organisms from other extreme habitats,[8] such as oceanic trenches.

Let us take a moment to contemplate this. Lake Vostok has been completely isolated for millions of years, and there is no way organisms from oceanic trenches could have migrated to it. Organisms adapted to Lake Vostok's extreme conditions are also unlikely to have always lived in it in their current form, seeing as Lake Vostok was once a "normal" surface lake. This leaves us with but one satisfactory explanation: as Lake Vostok transitioned into its present subglacial state, some of its "normal" organisms mutated to adapt to its challenges. The mutations they acquired, however, were far from random. In fact, they mirrored the mutations of organisms in similar, but faraway habitats, implying that every time nature rewinds the creation of this sort of extreme habitat, organisms acquire the same mutations.

Humans, too, are prone to acquiring nonrandom mutations. In Chapter 2 we learned that many people of sub-Saharan descent possess a mutation in the β-globin gene that causes red blood cells to collapse into a sickle-like shape, thus providing protection from the malaria-causing *Plasmodium* parasite. Like all mutations, this sickling mutation had long been thought to have arisen randomly. Indeed, it was once widely accepted that the mutation pops up at equal rates in populations all around the world, and only spreads in the ones inhabiting malaria-endemic areas. How can this hypothesis be tested? Comparing the number of mutation carriers found in different populations is unhelpful, because natural selection is expected to multiply the number of carriers in malaria-stricken regions, even if the mutation occurs very infrequently.

So how can we check how often the mutation occurs *before* it becomes visible to selective forces? One fascinating study solved this problem by looking for the sickling mutation in sperm produced by Africans and Europeans. This is an ideal way to measure "pure" mutation rates, because hemoglobin mutations that occur in sperm will receive no attention from natural selection until the commencement of hemoglobin production, long after fertilization. Consequently, if the sickling mutation arises randomly, it should be present in the sperm of Europeans as often as it is present in the sperm of Africans. But when researchers analyzed hundreds of millions of sperm cells, they discovered that the sickling mutation is generated at a much higher frequency in the sperm of Africans!

This is an extraordinary finding, but does it really mean that this mutation is nonrandom? Not necessarily. If Africans have higher mutation rates than Europeans, the mutation's preferential appearance in the sperm of Africans can be dismissed as an artifact. To rule out this possibility, the researchers looked into the mutation rate of the δ-globin gene, which is located near the β-globin gene, where the sickling mutation occurs. The δ-globin gene, it turns out, does *not* mutate faster in the sperm of Africans,[9] indicating that the abnormally high mutation rate in Africans is limited to a small stretch of the β-globin gene. This suggests that the sickling mutation arises in response to a need for malaria resistance, rendering it very nonrandom indeed.

Human genes, as we've seen, can pick and choose mutations to ward off *Plasmodium*. Can *Plasmodium* do the same to fight back? To answer this intriguing question, we must first explain how *Plasmodium* escapes detection by its greatest foe – the human immune system. Our immune system cannot peer into cells; it can only detect *extracellular* foreign molecules, for which it swiftly creates specific antibodies. Consequently, when *Plasmodium* enters the human bloodstream, it quickly invades red blood cells and hides within them.

Hunkering down in red blood cells, however, is not the parasite's ultimate goal; it also causes infected red blood cells to clump together and accumulate in many organs, a process associated with severe malaria.[10] To achieve this clumping, the parasite inserts special adhesive proteins into the membrane of each infected red blood cell. This is somewhat counterproductive, because now infected cells are studded with parasitic proteins, which the immune system can easily detect. To solve the problem it has itself created, *Plasmodium* resorts to engineering mutations in the approximately sixty genes encoding its adhesive proteins. These genes swap parts of their sequences with great alacrity, resulting in the appearance of a million or so novel mosaic proteins every two days.[11] The immune system, which is constantly producing specific antibodies to combat the parasitic proteins it has encountered, is thus kept busy churning out antibodies that are rapidly rendered obsolete. Needless to say, no other region of the Plasmodium genome is characterized by such a high mutation rate, because no other region would benefit from it. This strikingly nonrandom mutational mechanism is only used where it is needed.

The immune system, meanwhile, also employs nonrandom mutations to stay relevant. Remember that following our genome's replication, the DNA mismatch repair system identifies incorrect nucleotide pairings and snips out mutations before patching up the gaps with the "right" nucleotides. For most cells, this process is key to the preservation of protein function. But for the immune cells that produce antibodies, this process would eliminate any variation in antibody gene sequences, which, as we've seen, must encode an ever-changing array of antibodies. So how do these immune cells keep up with the diverse pathogens that invade our bodies? When their DNA mismatch repair system needs to patch up gaps in antibody genes, it often recruits a special error-prone DNA polymerase.[12] This trick ensures that our antibody genes are in a constant state of nonrandom mutation generation, or *mutagenesis*, which results in much-needed antibody variation.

Unlike the human immune system, which consists of almost two trillion cells,[13] bacterial immune systems are small and simple, though by no means ineffective. One of these systems, *CRISPR-Cas*, has risen to fame in recent years due to its usefulness in biotechnology. But the hype surrounding CRISPR-Cas has obscured the undeniable fact that this system is, at its core, a way for around half of all bacterial species to intentionally generate mutations.

Let us start at the beginning. For bacteria, life is all about contending with nonstop viral infections. Viruses, you see, are always bombarding the bacterial cells in which they seek to replicate. Fortunately, bacteria that employ the CRISPR-Cas system use it to capture short sequences of invading viral DNA and file them in a viral catalog of sorts. Suspicious DNA is always compared to this catalog, and if it matches any of the known viral sequences, it is destroyed before it can be replicated.[14]

What does this clever mechanism have to do with nonrandom mutations? The catalog, as it turns out, is none other than a segment of the bacterial genome, into which many snippets of viral DNA are inserted. In other words, many bacteria regularly induce insertion mutations in a specific region of their DNA. If mutations are supposed to be random, it seems bacteria didn't get the memo.

Thanks to the CRISPR-Cas system, many bacterial genomes are constantly expanding to include more viral sequences. This flies in the face of how most of us picture the genome – as a largely stable, unchanging mass of DNA that is only occasionally altered by mutations. The problem is that although this notion simplifies the definition of mutations, organisms typically have much more dynamic genomes. Let us consider the human genome. Every human starts out as a single fertilized egg cell, or *zygote*. An adult's genome, however, is in no way identical to a zygote's genome. That's because an adult doesn't have *a* genome, but rather trillions of genomes. To create an adult body, the zygote must divide into two cells, each of which divides into two cells, and so on, yielding the trillions of cells that make up an adult's body. Every division is preceded by DNA replication, which, as we've seen, is always accompanied by mutations. Thus, the human body develops into a mosaic of cells with different genomes, none of which are identical to the zygote's genome.

Bacterial genomes are even more dynamic. As humans, we are accustomed to thinking of DNA as something that is passed down generations, from parent to offspring, or from dividing cell to newly created cell. Bacteria pass DNA down to their descendants, but they also pass DNA to unrelated individuals, or even individuals from other species! *Horizontal gene transfer*, or *HGT*, as this phenomenon is known, is actually common in everything from fungi[15] to animals,[16] though we will focus on its implications for the bacterial genome.

Let us start with the basics. The bacterial genome typically consists of one chromosome (a large, circular DNA molecule), and multiple *plasmids* (little circular DNA molecules which often contain useful genes, such as ones associated with antibiotic resistance). When bacteria swap DNA via HGT, they do so by releasing or snatching up plasmids, resulting in genomes with ever-changing plasmid compositions. And what is the addition of a plasmid to the genome if not a major mutation?

Bacteriologists may object to this definition, because unlike the bacterial chromosome, plasmids are lost as readily as they are acquired,[17] rendering them merely temporary components of the genome. But who says that mutations must occur in a permanent part of the genome to be considered mutations? And what does "permanent" mean, anyway? Chromosomes, too, are sometimes lost.[18] Does that mean that all of the mutations that had occurred on them prior to their disappearance from the genome are retroactively nullified? And what about the chromosomes' disappearance? Is that not a colossal mutation in itself? Plasmids contain genes that play a key role in bacterial survival, and their exchange enables accelerated evolution,[19] since acquiring a gene via HGT takes considerably less time than evolving it from scratch. It is therefore only fair that we call HGTs mutations.

Objectors to the classification of HGTs as mutations should consider that HGTs may have once made up the vast majority of mutations. The genomes of the earliest cells, you see, were probably just collections of loose genes – a far cry from our modern multi-gene chromosomes. It has been proposed that back then, HGT rates were extremely high, resulting in tremendously unstable genome compositions. Consequently, species, which are defined by distinct genomes, could not exist in this primitive world. In the absence of separate species, evolution was a *communal* process undergone by populations of cells, rather than by individual lineages. According to this riveting hypothesis, HGT rates eventually plummeted, finally allowing separate species with distinct genomes to make their debut in the natural world.[20]

Our reluctance to see HGTs for the mutations they are may have something to do with how unfamiliar the concept of extrachromosomal genes is to many people. This is unfortunate, because bacteria are far from the only organisms with extrachromosomal DNA circles. The yeast genome, for example, consists of linear chromosomes[21] and many little DNA circles, whose accumulation causes aging.[22] And as it turns out, even mammalian cells can contain extrachromosomal DNA circles. Some of these circles are plasmids from infecting microbes, and some are formed when small stretches of DNA are pinched off the ends of chromosomes and circularized. To keep these mischievous DNA circles away from their precious chromosomes, mammals regularly deposit them in a cellular compartment aptly named the *exclusome.*[23]

Having established the ubiquity of extrachromosomal genes, we now turn to the question of whether the mutations known as HGTs are random. To answer this question, let us take a minute to think about what we would expect truly random HGTs to look like, before exploring whether reality lives up to our expectations. We would, for instance, expect plasmids to be "selected" at random for HGT. In other words, since each bacterium possesses multiple plasmids, for HGTs to be indisputably random, each HGT occurrence would have to entail one of these plasmids being singled out for transfer to another bacterium for no particular reason. In practice, however, many plasmids encode long, tube-like bacterial appendages called *pili* that latch on to other bacteria and transfer the pili-encoding plasmid to them.[24] This is nothing short of extraordinary, because it means that when a pili-encoding plasmid exists in one bacterium, it causes the bacterium to grow pili that transfer the plasmid to neighboring bacteria, which then grow pili and transfer the plasmid to *their* neighboring bacteria, and so on. Thus, the presence of a plasmid sets off a chain reaction of nonrandom mutations that introduce it into an ever-increasing number of genomes.

Let us set aside whole plasmids and zoom in on the genes they contain. If HGTs were truly random, we would expect bacteria to pass around a completely random selection of genes. Is this what happens in practice? Not at all. The delightfully named *housekeeping genes* – those responsible for maintaining basic cellular function – are much more likely to be horizontally transferred than the *informational genes* involved in transcription and translation.[25]

The genes that are horizontally transferred, then, are "chosen" in a nonrandom manner, but surely the order in which they are transferred is random? The hypothesis that genes are horizontally transferred in no particular order is actually very easy to test for two reasons. First of all, unlike many evolutionary phenomena, HGT occurs extremely frequently, enabling scientists to witness multiple instances of HGT in a relatively short period of time. Secondly, HGT is hard to miss because it leads to whole genes springing up in a genome. So what does research on this topic reveal? The order in which genes are horizontally transferred, it turns out, is so nonrandom that it is possible to predict which gene a bacterium will acquire next simply by checking which genes were already horizontally transferred to it.[26]

What else would we expect if these mutations were random? To be random, HGTs would have to occur at random rates. Their rates, however, are profoundly influenced by environmental conditions. This was demonstrated by a study that looked into the gut bacteria of fifteen different human populations. When researchers sequenced and compared the genomes of gut bacteria from each population, they came to the surprising conclusion that industrialization and urbanization are associated with higher HGT rates.[27] So the rates of HGTs are nonrandom, but are these intriguing mutations random in other ways? If, for example, multiple genes are horizontally transferred together, is this at least a coincidence?

Remarkably, genes are often horizontally transferred together in an outrageously uncoincidental manner. *Mesorhizobium ciceri*, for instance, is a nitrogen-"fixing" bacterium, which means that it converts atmospheric nitrogen into ammonia.

The bacterium can leverage this unique ability to form a symbiotic relationship with a legume plant, which offers it an endless supply of fixed carbon in exchange for ammonia. But in order to transition to a symbiotic state, an *M. ciceri* bacterium must first receive a very specific HGT from another *M. ciceri* bacterium that is already in a symbiotic state. First, three separate chromosomal regions of the symbiotic bacterium are excised and joined together. The resulting DNA molecule then undergoes circularization and is horizontally transferred to a neighboring bacterium. There, it disassembles into its original three parts, each of which integrates into a specific site in the recipient genome. Upon integration, the recipient cell gains the ability to forge a symbiotic relationship with legumes,[28] as well as the ability to initiate this marvelously nonrandom type of HGT.

HGT is not limited to neighboring cells; it can also take place between two genomes inhabiting the same cell. This might sound outlandish, since we are accustomed to thinking of a cell as a container for a single genome. But the truth is that the cells that make up our bodies can contain quite a few genomes because they house many *mitochondria*. What are mitochondria, you ask? To answer this question, we must first understand what they used to be.

Mitochondria are the descendants of a bacterial cell that was internalized by a much larger cell over 1.45 billion years ago.[29] The reason for this internalization is up for debate. Perhaps the large cell gobbled up the bacterium and failed to digest it. Alternatively, the bacterium may have practiced a parasitic lifestyle which consisted of invading larger cells and living off their resources. Either way, the large cell ended up with an intact bacterium living within the confines of its membrane. It was then that something unexpected occurred. The large cell "discovered" that the bacterium could produce energy very efficiently, so it began to borrow the bacterium's energy. Meanwhile, the bacterium "discovered" that if it provided the large cell with energy, the large cell would happily attend to all its other needs. So the bacterium gradually stopped producing most of the proteins that had allowed it to live independently and evolved an ever-increasing reliance on the large cell. Over many generations, it degenerated into the energy-producing organelle we now call the *mitochondrion*. Our cells, which are the descendants of the large cell, contain multiple mitochondria apiece. And although little remains of the bacterium that once was, each mitochondrion still contains a small, vestigial bacterial genome.[30]

Mitochondrial DNA, as this vestigial bacterial genome is known, is largely confined to our mitochondria. Mitochondrial genes do, however, occasionally jump out of mitochondria and into the nucleus, where the cellular genome is stored. HGT, then, can take place between the mitochondrial and nuclear genomes of the same cell. But that's not all – following this bizarre HGT, mitochondrial genes integrate themselves into cellular DNA, becoming *nuclear mitochondrial sequences*, or *NUMTs*.

NUMTs are, by definition, the result of insertion mutations, and at this point it should come as no surprise that these mutations are anything but random. This nonrandomness can be easily spotted when one examines the distribution of NUMTs in the nuclear genome. NUMTs, it seems, have a preference for insertion into regions of the DNA molecule that are characterized by high curvature, as well as sequences containing many A and T nucleotides.[31]

It should also be noted that although NUMT insertions were once thought to be rare events, evolutionarily speaking, recent studies have revealed that these mutations occur repeatedly over a human lifetime, especially in various parts of the brain. And the rate at which NUMT insertions occur is also nonrandom. One study found that while cells grown in a laboratory setting accumulate just a handful of NUMTs every month, mitochondria respond to stressful conditions by releasing more snippets of their DNA, which can accelerate the NUMT insertion rate by up to five times![32]

Genomes, it is clear, are adept at giving each other genes in a decidedly nonrandom fashion. But can a genome gift *itself* genes? And could such a weird mutation be nonrandom as well?

Before delving into this mystifying subject, let us recall how mutations "normally" work. An enzyme called *RNA polymerase* usually transcribes a gene from DNA into RNA, which is later translated into a protein. So if a mutation changes the DNA sequence, RNA polymerase will unwittingly produce a mutated RNA sequence, which will then be unwittingly translated into a mutant protein.

Some bacteria, however, have little patience for traditional genetics when faced with viral infection. One groundbreaking study found that these anomalous organisms possess an enzyme called *reverse transcriptase*, which transcribes RNA back into DNA. In an uninfected bacterium, this reverse transcription produces single-stranded extrachromosomal genes that don't seem to do much. They are, in other words, apparently inconsequential genes that the genome has added to itself. But if the bacterium is attacked by viruses, a second strand is quickly added to the extrachromosomal genes, which tend to accumulate quickly within infected bacterial cells. At this point, the extrachromosomal genes are transcribed into RNA and then translated into proteins that seem to inhibit cell growth and induce dormancy, thus thwarting the viruses' plans of reproducing and moving on to neighboring bacteria.[33]

It is hard to overstate how astonishing this kind of mutation is. Genetic information, as we've seen, usually follows this very simple path:

$$\text{DNA} \rightarrow \text{RNA} \rightarrow \text{protein}$$

Crucially, once the genetic information "leaves" DNA, it cannot alter it in any way. Gene expression, in short, does not normally lead to mutation. But this is not the case in the sensational defensive mechanism described earlier, which can be summarized as follows:

$$\text{DNA} \rightarrow \text{RNA} \rightarrow \text{DNA} \Longrightarrow \text{DNA} \rightarrow \text{RNA} \rightarrow \text{protein}$$

Here, gene expression leads to the expansion of the genome via some breathtakingly convoluted mutations. And since these mutations arise in response to viral infection, they can lay no claim to randomness. In other words, although the idea that mutations are random has reigned supreme since the dawn of neo-Darwinism, unmistakably nonrandom mutations such as these force us to rethink the principles of evolution.

CHAPTER 5

Taxonomic Tangles

Mutations, as we've seen, are often nonrandom. But is mutational nonrandomness meaningful? Or is it just an insignificant curiosity that, although interesting, ultimately has very little impact on life? To answer these questions, let us first familiarize ourselves with one of the most-used diagrams in biology: *evolutionary trees.*

Evolutionary trees, also known as *phylogenetic trees*, are rough graphical representations of how species, or families of species, diverged from each other. An evolutionary tree will typically depict a limited number of species branching off from a known or unknown ancestral species. This, of course, is only a small part of the big picture, because if we assume that all organisms are descended from a single, primordial ancestor, then drawing a full evolutionary tree will require adding branches for *all* species, extant and extinct.

Darwin himself was an evolutionary tree sketcher. To make sense of the natural world, he recorded various observations and ideas in a series of notebooks that helped him develop his theory of natural selection over the course of many years. In 1837, he drew a simple evolutionary tree in one of his notebooks. Tucked away between accounts of conversations Darwin had with his barber and insights gleaned from his visits to the zoo, this evolutionary tree is easy to miss. But its depiction of new species branching off from ancient ones reveals that at the time of its drawing Darwin had come to the momentous conclusion that all living things are related. The two words scrawled above the tree – "I think" – go one step forward, capturing Darwin's uncertainty as he struggled with the enormous implications of his theory.

Evolutionary trees have come a long way since Darwin's rudimentary sketch. The evolutionary trees biologists now use can be *rooted*, in which case the most recent common ancestor of the depicted species is included in the tree, or *unrooted*, in which case related species are depicted, but no assumptions about ancestry are made.[1] Similarly, evolutionary trees can be *phylograms*, which means that their branch lengths are proportional to the species' divergence in features, or *chronograms*, which means that their branch lengths are proportional to the time that has passed since divergence.[2] Alternatively, they can be *cladograms*, which means that their branch lengths represent nothing at all.[3]

The way evolutionary trees are constructed has also, well, evolved. In Darwin's day, evolutionary trees were based on observable similarities between species, such as wing size or skull shape. But nowadays, the DNA of different species can be readily sequenced and compared, providing evolutionary biologists with unprecedented amounts of data from which highly accurate evolutionary trees can be derived. At least, that's how harnessing sequencing technologies to construct evolutionary trees should work. In practice, however, quite a few attempts to construct sequence-based evolutionary *trees* have yielded ambiguous *bushes* instead.

How is this possible? To understand the origins of this unexpected outcome, we must first examine how genomes change after diverging. In most cases, the genomes of distinct species will accumulate distinct mutations, simply because they are evolving separately. This makes for a very simple rule of thumb: the more similar the genomes of two species, the more closely related they are, because species that diverged long ago should have very different genomes by now. But sometimes, the same mutation occurs at the same genomic position in two (or more) distinct species. This phenomenon, which is known as *homoplasy*, results in a very confusing state of affairs: species share a mutation, but not because they inherited it from a common ancestor.

You may think that homoplastic mutations would be extremely rare, but they actually arise surprisingly frequently. In fact, homoplastic mutations are so common that they can render the DNA sequences of distant species similar enough to yield completely erroneous evolutionary trees. Think of it this way: to assess divergence by comparing DNA sequences, one requires many random mutations, which enable genomes to become sufficiently dissimilar. But if enough homoplastic mutations conspire to increase the similarity of long-diverged genomes, the sequencing of different genes can result in alternative evolutionary trees.

Take, for example, the human-chimpanzee-gorilla evolutionary tree. Chimpanzees have long been considered our closest relatives,[4] which means that we diverged from chimpanzees *after* our common ancestor diverged from gorillas. However, sequencing the same set of genes in humans, chimpanzees, and gorillas, and comparing the sequences of each gene to construct an evolutionary tree has produced unexpected results. The sequences of 55% of the genes yield the traditional evolutionary tree, in which humans and chimpanzees are most closely related. But thanks to homoplastic mutations, the sequences of 18% of the genes "prove" that gorillas are our closest relatives, while 20%* of the genes "prove" that chimpanzees and gorillas are more closely related to each other than they are to humans!

To be fair, a certain number of homoplastic mutations is inevitable. Just as two colleagues will occasionally turn up to the office wearing the same outfit, so will the same mutation occasionally pop up in two species. But does it make statistical sense for homoplastic mutations to arise frequently enough to generate false evolutionary trees? Researchers studying ambiguous evolutionary bushes resembling the human-chimpanzee-gorilla one have calculated that if homoplastic amino acid substitution

* The sequences of the remaining genes are uninformative.

mutations occurred randomly, they would make up 2–5% of all amino acid substitutions. In practice, though, homoplastic mutations make up an astounding 10–15% of amino acid substitutions,[5] demonstrating their deeply nonrandom nature. Homoplastic mutations are so nonrandom, in fact, that DNA sequencing is now seen as a largely ineffective tool for evolutionary tree construction[6] – an astonishing fact that speaks volumes about the impact of nonrandom mutations.

Evolutionary trees, incidentally, are also sometimes hard to construct for certain extinct species. To understand why this is the case, consider the problem of pinpointing when species became extinct. This is usually a simple task: by identifying the last era in which a given species appears in the fossil record, one can deduce that it subsequently underwent extinction. However, this method is not bulletproof, because some "extinct" species later reappear in the fossil record following gaps of up to millions of years. A few of these bewildering species, in fact, are currently extant!

How can we make sense of the *Lazarus effect*, as this intriguing phenomenon is called? One possible explanation is that since the fossil record is infamously incomplete, the Lazarus effect is purely accidental.[7] Very few organisms are fossilized after death, and only a tiny fraction of those that *are* fossilized end up being discovered by paleontologists. It is therefore only natural that some species will "fail" to produce fossils for part of their existence, resulting in their apparent extinction and revival. The Lazarus effect may also be caused by a massive reduction in population size. In other words, if a population is on the verge of extinction, it will probably "fail" to produce fossils for purely statistical reasons; but if it later recovers, it will reappear in the fossil record.[8]

The Lazarus effect can be reproduced in laboratory settings by inducing a bacterial population crash with lethal temperatures. If survivors emerge, it is due to the occurrence of "savior" mutations that enable survival, and later, population rebounding. But are these "savior" mutations similar to the mutations that arise under ordinary circumstances? And if not, are they random in any other way? To answer these questions, let us examine two similar studies involving the exposure of bacteria to unbearably high temperatures.

In the first study, 115 identical bacterial populations were exposed to the high, but not lethal, temperature of 42.2°C. A single bacterium from each population was then randomly selected for genome sequencing, revealing over 1,000 mutations that may have helped the bacteria adapt to life at 42.2°C. All of the populations acquired at least one mutation in a gene involved in transcription.[9]

In a second study, carried out by a different team of researchers, forty-four identical bacterial populations were exposed to the *lethal* temperature of 43.0°C. Although this experiment was repeated with new bacterial populations every week for seven weeks, very few populations experienced the Lazarus effect. And when the genomes of the few surviving bacteria were sequenced, two distinct categories of mutations emerged: mutations in transcription-related genes and mutations in genes responsible for dealing with damaged proteins. This indicates that "savior" mutations are not run-of-the-mill high-temperature adaptation mutations, which, as we've seen, tend to focus on transcription. Intriguingly, specific mutations in one of the categories tended to appear in multiple populations in the same week, while specific mutations in the other category appeared repeatedly *across* weeks, demonstrating that the two categories are characterized by different kinds of mutational nonrandomness. The sequencing results also revealed a bias towards one of the categories, which, incidentally, was itself biased towards frameshift mutations. Last but not least, none of the populations experiencing the Lazarus effect acquired mutations from *both* categories. This is, of course, a statistically improbable finding that adds yet another layer of nonrandomness to this enigmatic phenomenon.[10]

CHAPTER 6

Biased Bursts

Mutations, it is safe to say, are not random. Why is this the case? One of the most important reasons is that the mutation *rate*, or the frequency with which mutations arise, is not random. There are several ways to calculate mutation rate, including mutations per nucleotide per cell division, mutations per gene per generation, and mutations per genome per generation. All of these, however, are ultimately determined by a dynamic balance between opposing evolutionary forces. To shed light on these shadowy forces, one might ask why a high mutation rate would be advantageous, and contrarily, why a low mutation rate would be advantageous. A high mutation rate provides more opportunities for adaptive mutations to occur – an undoubtedly positive outcome. Unfortunately, it also provides more opportunities for deleterious mutations to occur, meaning that a low mutation rate is often preferable. This can only be achieved, however, by investing considerable resources in proofreading DNA and correcting mutations, and consequently, such budgetary considerations determine how much the mutation rate can be reduced.

The tug-of-war between these evolutionary forces ends up shaping the mutation rate in surprisingly nonrandom ways, as we shall soon see. But first, what do we mean when we say that the mutation rate is nonrandom? For starters, different species are characterized by very different mutation rates. In some viruses, for example, a mutation will arise roughly once every generation, whereas some microbes will only see a mutation arise about once every 300 generations.[1] Similarly, the rate of primate substitution mutations is highest in New World monkeys, lower in Old World monkeys, lower still in Great Apes,[2] and lowest of all in humans,[3] where it may still be falling.[4] Now, if mutations were random, wouldn't we expect them to arise in humans as often as they arise in our New World monkey cousins?

Different species don't just happen to possess different mutation rates; there are several explanations for this glaring disparity. The first is that although DNA replication is by and large an accurate operation, just how accurate it is can vary greatly between species. Some species employ stringent enzymes to replicate, proofread, and correct their genomes, while others employ less efficient enzymes to carry out these tasks. Curiously enough, DNA replication accuracy is strongly correlated with genome size. The reason for this correlation seems to be that large genomes tend to produce complex systems, which are particularly sensitive to deleterious mutations.[5] This would suggest that as our ancestors evolved larger genomes, they were forced to take steps to reduce their mutation rates, resulting in an extremely low human mutation rate.

Species also differ greatly in cell division times and generation times, which means that we can expect species that replicate their genomes faster to undergo more replications, and therefore accumulate more mutations, per unit of time. And an additional species-specific characteristic – metabolic rate – contributes to the mutation rate disparity because higher metabolic rates generate more mutagens.[6] Similarly, the different habitats of different species play a major role in determining their dissimilar mutation rates. One study found that in high-energy regions, such as those characterized by high UV radiation, species are characterized by relatively high mutation rates.[7]

Speaking of how environmental conditions determine mutation rates, the *Chironomus riparius* midge, which inhabits everything from cold northern regions to sweltering southern ones, mutates at a temperature-dependent rate. Moderate weather, it turns out, results in moderate mutation rates for midges, while extremely hot and extremely cold climates significantly increase mutation rates.[8] Bacteria, too, respond to stressful environmental changes by elevating their mutation rates, thus producing a higher supply of mutations which could potentially help them adapt to a dynamic environment.[9] Naturally, once adaptation is achieved, mutation rates decline again.[10]

This intentional mutation rate boosting has been observed in bacteria suffering from high population densities or extreme starvation,[11] as well as bacteria forced to endure the presence of viruses.[12] It has also been observed in bacteria exposed to a range of antibiotics, especially at low concentrations. Accelerated mutagenesis, in fact, is at the heart of the emergence of *superbugs*, or bacterial strains that are resistant to multiple types of antibiotics. Many patients, you see, will stop taking their prescribed antibiotics when they start to feel better, instead of completing the full course. Rather than eradicating all of the pathogenic bacteria in their bodies, the noncompliant patients allow a few to survive in a sublethal, diminishing concentration of antibiotics. Under these conditions, elevated mutation rates allow bacteria to "try out" various resistance strategies, culminating in the acquisition of resistance to multiple types of antibiotics, which renders our pharmaceutical arsenal largely ineffective.[13]

But how do bacteria increase their mutation rates in the first place? The answer is that although they usually employ accurate DNA polymerases, bacteria also possess genes that encode error-prone DNA polymerases. Normally, these inefficient proteins are passed over in favor of their overachieving counterparts, but when bacteria are subjected to stressful conditions, they allow the error-prone replicators to take over and litter the genome with mutations.[14]

A close examination of the ongoing interactions between parasites and their hosts reveals that parasites, too, influence mutation rates. To understand why this is the case, ask yourself: where do organisms obtain their parasites? Some are undoubtedly acquired by happenstance, but a great deal are passed down from parents to offspring. The malaria-causing *Plasmodium* parasites, for instance, can infect an individual who happens to be bitten by an infected mosquito. But if they find themselves in the blood of a pregnant woman, they are very likely to cross the placenta and infect the fetus.[15] Moreover, in species characterized by post-hatching or post-birth parental care, offspring continue to be exposed to their parents' parasites right up to the cusp of adulthood. This is significant because it means that while individuals may occasionally be exposed to a stranger's parasites, they are extremely likely to encounter the same parasites that plague their parents, and should perhaps be expected to. It is therefore unsurprising that high parental parasite transmission is associated with high mutation rates, which give hosts a leg up in the multigenerational battle against parasites.[16] Mutations, then, aren't random occurrences, but weapons wielded deliberately to stave off parasites.

Mutation rates don't just depend on external conditions; they also depend on the cellular location of DNA, as exemplified by the mitochondrial DNA mutation rate being approximately twenty times higher than the nuclear DNA mutation rate.[17] And what, exactly, is the nuclear DNA mutation rate? This question is impossible to answer, because the nuclear mutation rate depends on cell type. A good example of this phenomenon can be observed by comparing germ cells (the precursors of sperm and egg cells) with somatic cells. The mutation rate of somatic cells is roughly two orders of magnitude higher than that of germ cells,[18] suggesting a nonrandom mechanism for either boosting somatic mutations or suppressing their germline counterparts.

But that's not all – in any given cell type, different regions of the genome are characterized by different mutation rates. Although this may seem counterintuitive, the frequency of DNA damage varies in different regions of the genome, as does the frequency of DNA repair.[19] Consequently, some genomic regions mutate rapidly, some mutate at a moderate rate, and others barely mutate at all. These differences are evident when one compares the mutation rates of different chromosomes, but they also exist at a much smaller scale – adjacent nucleotides, for example, have been shown to mutate at wildly different rates.[20]

If mutation rates are ultimately determined by local DNA damage and repair frequencies, doesn't that mean that mutations are random, albeit in an indirect way? Perhaps, but only if every segment of the genome is randomly "assigned" DNA damage and repair frequencies. This, however, does not seem to be the case. In the unicellular *Tetrahymena thermophila*, for instance, genes are characterized by a significantly higher mutation rate than intergenic regions.[21] Similarly, the mutation rates of the most highly expressed genes in yeast are roughly twice as high as the mutation rates of genes that are mostly inactive.[22]

Some genes avoid mutating at all costs. When researchers sequenced and compared the genomes of hundreds of *Arabidopsis thaliana* (thale cress) plants, they discovered an overrepresentation of essential genes in regions marked by extremely low mutation rates.[23] *A. thaliana*, it is clear, actively shields its most important genes from mutations in order to prevent the disruption of critical cellular processes. The unassuming little plant does, however, allow inessential genes to mutate, presumably because it can afford to lose at least some of their functionality in its quest to evolve new adaptations. Now, if *A. thaliana* can pick and choose the types of genes subjected to mutations, how can mutations be said to arise randomly with respect to their effects on organisms?

In humans, mutation rates vary considerably throughout the genome – with important implications for our evolution. This is exemplified by some particularly rapidly mutating genomic regions known as *human accelerated regions*, or *HARs*. When researchers looked into the role that just five HARs played in human brain evolution, they came to the unexpected conclusion that these HARs may be responsible for the evolution of the higher cognitive functions that set us apart from other primates.[24]

But high mutation rates are not content to drive the acquisition of complex linguistic capabilities and other advanced traits. They also play a much more profound role in evolution by determining how quickly mutations catch on. New mutations, you see, don't always spread in populations. When they do, and when the mutated versions of genes become the new default in a population, we say that the mutations underwent *fixation*. The occurrence of a mutation by itself cannot produce real evolutionary change; it is only the fixation of a mutation that can alter life at a macro level. But what does fixation have to do with high mutation rates? This is where things get interesting. When mutations arise very often, a larger percentage of them tend to undergo fixation.[25] Put simply, the more a gene is bombarded with mutations, the less picky it becomes about accepting them. So when mutation rates are nonrandomly elevated in certain genomic regions, the resulting mutations are nonrandomly favored for fixation.

The mutation rates of different genomic regions, it should be noted, are not set in stone. In fact, mutation rates will often change over the course of an organism's lifetime. In humans, hundreds of mutations accumulate quickly immediately after conception. Although this process later tapers off, research on brain development has revealed that mutation rates accelerate once again during the brain-building *neurogenesis* stage, before plummeting to the relatively low rates characteristic of adulthood.[26] Similarly, *E. coli* bacteria mutation rates start out strong and then slow down in aging bacterial colonies.[27]

The ebb and flow of mutation rates over a lifetime pales in comparison to the fluctuation of mutation rates over a species' *evolution*. The human mutation rate, for example, appears to have increased significantly in the European population following its migration out of Africa roughly 200 millennia ago. This increase is hard to miss when one investigates the mutations underlying *melanoma*, a skin cancer caused by excessive exposure to UV radiation from the sun. Melanoma is virtually unheard of in humans of African descent, because dark skin is the result of a pigment called *melanin*, which blocks UV rays very efficiently. But not all light-skinned, and therefore melanin-poor, humans are at an equal risk of developing melanoma. Light-skinned Asians, for instance, are nowhere near as likely as Europeans to suffer from melanoma. So what is it about Europeans that makes them so susceptible to sunlight?

This brings us back to the elevated mutation rate acquired by Europeans after venturing out of Africa. The vast majority of melanoma mutations are substitutions of C nucleotides with T nucleotides, and almost a third of these replace a TCC triplet with a TTC triplet. One intriguing study revealed that when the European mutation rate increased, the rate of TCC-to-TTC mutations skyrocketed, as did the rates of several other C-to-T mutations.[28] Thus, a blatantly nonrandom mutation rate led to the occurrence of certain mutations in just one light-skinned population, resulting in a very nonrandom global distribution of melanoma cases.

CHAPTER 7

Strategic Spots

The subject of nonrandom mutation rates overlaps, to an extent, with the equally intriguing subject of nonrandom mutation *locations*. That's because when it comes to *where* mutations occur, not all locations – anatomical, genomic, or spatial – are created equal. Take the mutations caused by exposure to mutagens. Imagining an organism as a blob of cells that are all equally affected by exposure to, say, radiation may simplify things, but mutagens tend to affect different tissues in very different ways. UV rays, for example, wreak havoc on our skin, but have little to no effect on internal organs, which are not exposed to sunlight. One enlightening study on the subject revealed that the skin cells found in the eyelids of middle-aged and elderly humans contain more than their fair share of mutations. In fact, decades of absorbing UV rays result in up to 180 mutated genes per eyelid skin cell[1] – significantly more than the number of genes expected to mutate by chance.

Skin cells are not the only cells where nonrandom mutagenesis occurs; this peculiar process is also evident in our germ cells. As we saw in Chapter 3, germline mutations are the only mutations that really matter, evolutionarily speaking. Mutations that arise in somatic cells can have far-reaching effects on an individual, of course, but for a mutation to make a dent in an evolutionary process, it must first escape its body of origin. And since the mutations in our germ cells are the only ones capable of pulling off intergenerational transmission, the manner in which they occur is of special interest to us. So how do the mutations in germ cells differ from the mutations that arise elsewhere? For some reason, these potentially game-changing mutations are much more clustered than we would expect them to be if their occurrence were truly random.[2] Something, it is clear, is propelling these mutations into existence in the vicinity of other mutations.

It's not just specific cell types that boast mutational quirks. Even the two strands of the DNA double helix are characterized by wildly different mutational processes![3] One reason for this seemingly bizarre difference between strands is that genes are encoded on just one DNA strand. This means that when two strands are separated to allow a gene's transcription, the transcribed strand is quickly covered, and hence protected, by RNA, whereas the nontranscribed strand's unpaired nucleotides are exposed to a host of mutagenic factors. Consequently, mutations occur primarily on nontranscribed strands of DNA. This was demonstrated in one fascinating study on G/C-to-A/T mutations, which are the most common nucleotide substitution mutations in many species. When the study's authors looked into the origins of these mutations in *E. coli*, they discovered that a whopping 77% of them start out on the nontranscribed strand,[4] highlighting yet another aspect of mutational nonrandomness.

Transcribability is just one criterion by which we classify DNA strands. When DNA is replicated, the two strands are separated, and each one serves as a template for the construction of a new strand. The two original strands, however, do not undergo identical processes of replication. One strand, known as the *leading strand*, is replicated continuously from end to end, while the second strand, or *lagging strand*, is replicated in fits and starts.

To understand what the distinction between leading and lagging strands has to do with nonrandom mutations, we must first take a closer look at how and when DNA replication takes place. In bacteria, replication and transcription often take place simultaneously. This means that as DNA polymerase slides along a strand of DNA, building a complementary strand, RNA polymerase can slide along the *same strand*, transcribing it into RNA. The leading strand features codirectional replication and transcription, which precludes any interference between the two processes. Replication and transcription of the lagging strand, however, progress in opposite directions, yielding head-on collisions between replication proteins and transcription proteins. Naturally, these collisions cause DNA polymerase, which normally scans strands very carefully, to lose its place in the sequence. When it resumes replicating, DNA polymerase doesn't necessarily pick up where it left off, which is why collisions often result in mutations.

Collisions, then, can be expected to bias the lagging strand in favor of mutations, but do collisions occur frequently enough to produce a discernible effect? Absolutely. Not only are single-nucleotide mutations considerably more common in lagging strands, but many genes appear to have "fled" to the leading strand to escape the lagging strand's excessive mutagenesis. Consequently, in *Bacillus subtilis* bacteria, 75% of all genes and a staggering 94% of essential genes are found on the leading strand. But what about the lagging strand's few remaining genes? These genes, it turns out, are about half as long as the genes found on the leading strand.

To make sense of this disparity, let us put ourselves in the shoes of a lagging strand gene. How can this gene avoid the onslaught of potentially deleterious mutations that characterize the lagging strand? It can skip over to the leading strand, which, as we've seen, is what most lagging-strand genes have done throughout evolutionary history. But even if it stays behind on the lagging strand, it can still minimize its chances of hosting collisions by mutating into a shorter version of itself.[5] The logic is simple: the shorter the gene, the less room available for collisions.

Surprisingly, DNA replication does not require the collisions discussed earlier to produce nonrandom mutations. In yeast, for example, mutability can vary sixfold across a single chromosome, and is correlated with *replication timing*.[6] Interestingly, this correlation seems to be very common. Indeed, a good rule of thumb is that regions that are replicated early on, such as active genes found in the nuclear interior, mutate very little, whereas regions that are replicated at the very end of the DNA replication process, such as repetitive sequences found in the nuclear periphery, are marked by an increased susceptibility to mutations.[7]

Why should replication timing matter? This may seem puzzling at first, until one considers how the replication process changes over time. When the first genomic regions are replicated, there are plenty of free nucleotides for DNA polymerase to incorporate into a new, growing strand. But the number of nucleotides floating around decreases as replication progresses,[8] and DNA polymerase becomes increasingly likely to incorporate "wrong" nucleotides into a strand when the "right" nucleotides aren't readily available. Thus, the late-replicating regions get the short end of the mutational stick.

Incidentally, gene *distribution* is also correlated with replication timing. The mutations that give rise to new genes will be discussed at length in Chapter 16, but for now, let us examine the chromosomal positioning of these riveting mutations. If gene-creating mutations were truly random, we would expect to see genes scattered haphazardly along a chromosome. However, a study analyzing over 900 bacterial species revealed that roughly two thirds of bacterial gene families are arranged nonrandomly. Moreover, the study demonstrated that genes encoding transcription and translation proteins cluster in regions that undergo early replication,[9] indicating that our chromosomal architecture is not a product of chance.

As we saw in Chapter 6, different genomic regions – and even adjacent nucleotides – often vary greatly in mutability. The most elegant example of this phenomenon is probably *E. coli*'s circular chromosome, which features a striking wave-like mutability pattern around its circumference.[10] But the strongest evidence of nonrandom mutations is undoubtedly the existence of *mutation hotspots*, or genomic regions that are so prone to mutating, that their mutability exceeds that of neighboring regions by up to several orders of magnitude.[11] Now, are mutation hotspots randomly scattered throughout the genome? Of course not! Mutation hotspots tend to appear in regions that are crucial to adaptive evolution, such as those containing genes that encode proteins involved in detecting environmental stimuli or mounting an immune response.[12]

Moreover, many of the areas in which the DNA molecule folds into irregular three-dimensional structures are mutation hotspots.[13] But what, exactly, does it mean for DNA to fold into an irregular three-dimensional structure? The "canonical" DNA conformation is a right-handed double helix, or two strands twisted around each other in the same direction your fingers turn when you use your right hand to give a thumbs-up. The double helix can, however, assume a *left-handed* conformation, or give way to *triple-strand* or even *quadruple-strand* structures.

Non-canonical DNA conformations such as these are often mutation hotspots for two reasons. The first is that the segments linking canonical and non-canonical structures are subjected to a great deal of strain, which produces strand breaks and plenty of opportunities for mutations to occur when these breaks are repaired.[14] The second reason is that the non-canonical structures themselves are exceedingly difficult for the replication machinery to navigate, which means that single-stranded DNA is often exposed for extended periods of time, during which damage is likely to occur. Furthermore, the replication machinery is likely to fall off the oddly shaped DNA repeatedly, resulting in an error-prone replication process.[15] Needless to say, if mutability depends on factors such as gene function and DNA conformation, then mutagenesis can lay no claim to randomness.

Another hard-to-miss mutation hotspot is the chromosome's *centromere*. During DNA replication, each chromosome is replicated, yielding two identical chromosomes that are initially linked to each other at a region called the centromere, which divides each chromosome into two *arms*. When it is time to divvy up the chromosomes between the two nascent cells, special fibers attach to each chromosome's centromere and pull the pair apart, thus ensuring that each cell ends up with an identical set of chromosomes. Many centromeres are mutation hotspots,[16] which is significant because the mutations that tend to arise on centromeres are not run-of-the-mill mutations. Centromere mutations can induce chromosomes to swap whole chromosome arms, effectively undermining the cell's efforts to divide its gene copies equally between offspring.

Mutations have also been shown to depend on how DNA is packaged, opening up new avenues of nonrandomness. As we've seen, DNA is a (mostly) double-stranded chain. But how can a chain that is approximately two meters in length fit inside the nucleus of each human cell?[17] The answer is that DNA is packaged into a dense structure known as *chromatin*, which also prevents strand entanglement. That said, it stands to reason that if chromatin is *too* densely packed, the genes within it will be inaccessible to RNA polymerase, and will consequently never yield proteins. To solve this problem, at any given time certain genomic areas are packaged into very dense *closed chromatin*, while others are packaged into looser *open chromatin*. The two kinds of chromatin are marked by very different kinds of mutations. Nucleotide substitutions are much more common in closed chromatin, whereas insertions and deletions are a hallmark of open chromatin.[18]

But that's not all – chromatin structure also determines mutability at a deeper level. To begin packaging DNA into chromatin, the double helix is wound around protein complexes known as *histones* like yarn is wound around spools. Incredibly, the mutability of DNA between histones is up to twice as high as the mutability of DNA wrapped around histones,[19] suggesting that histones can provide some sort of protection from the occurrence of mutations. But histones are not randomly distributed throughout chromosomes; the DNA sequence plays a key role in determining which areas will be occupied by histones.[20] In other words, the DNA sequence dictates histone positioning, which, in turn, determines the mutability of the DNA sequence. Does that sound like a random process?

Zoom into the DNA sequence, and another class of mutation hotspots emerges – the *repetitive sequences*. DNA is often peppered with repetitive sequences of varying lengths, which can make up a sizable portion of genomes. In humans, for example, repetitive sequences, which play important structural roles in chromosomes, may constitute more than two thirds of the genome.[21] Not only are repetitive sequences and their neighboring nucleotides more likely to mutate,[22] but they also demonstrate a preference for certain kinds of mutations, such as deletions and frameshift mutations.[23]

At first glance, these two kinds of mutations seem to have very little in common, but a closer examination of repetitive sequence mutagenesis reveals why deletions and frameshift mutations are the obvious choice for this unique process. Imagine, if you will, DNA polymerase replicating a stretch of DNA containing two identical, but nonadjacent repetitive sequences. If DNA polymerase slips while scanning the first one, it may lose its place and resume replicating from the *second* one, resulting in the deletion of the segment between the two repetitive sequences. Similarly, since each repetitive sequence contains multiple copies of a mini-sequence, if DNA polymerase slips and resumes replicating from a different position within the same repetitive sequence, it can easily yield a frameshift mutation.

Certain genes also serve as nonrandom mutation hotspots. As we saw in Chapter 6, increasing the mutation rate is a common bacterial response to stress. The increased mutation rate is not a general one, because as we've seen, essential genes must be protected from mutations at all costs. Instead, in a brazen display of nonrandomness, *E. coli* bacteria respond to starvation by facilitating DNA breaks in certain genomic locations.[24] They then proceed to utilize an error-prone repair mechanism to patch up the DNA molecule,[25] effectively generating mutations on demand. Needless to say, the mutations produced by the faulty repair mechanism tend to cluster at the sites "selected" by *E. coli* for accelerated adaptation,[26] while other sites are unaffected.

Having established that mutation type and frequency are nonrandom with respect to anatomical/genomic location, we turn to a higher dimension of nonrandomness, which is derived from *epigenetics*. But to understand how epigenetics can influence mutations, we must first understand what this term means. The Greek word *epi* means "above", so *epigenetics* literally means "above genetics". What, you may be wondering, can we find above the genetic world of nucleotides and genes? The answer is that there are two ways to change DNA. The first is to physically alter its sequence by cutting the molecular chain and inserting, deleting, or replacing nucleotides. This, of course, is the definition of a mutation. The second, and perhaps more interesting, way to change DNA is to attach little molecules to nucleotides or to the histones around which DNA is wrapped. The attachment (or removal) of these little molecules, known as *epigenetic marks*, does not alter the nucleotide sequence, but nevertheless has an immense impact on its expression. And although many epigenetic marks are erased when DNA is passed down to offspring, providing each new generation with a clean slate of sorts, some epigenetic marks are inherited,[27] a fact that places them on equal evolutionary footing with bona fide mutations.

A good example of how epigenetic marks alter gene expression can be found in the crucial process of *cellular differentiation*. Every organism starts out as a single cell. We humans start out as a fertilized egg cell, or *zygote*, which divides repeatedly to give rise to the trillions of cells that make up a human body. In adults, cell division is largely unremarkable because a dividing cell will always give rise to identical, or at least very similar, cells. But back when we were embryos, our cell divisions were inextricably intertwined with cellular differentiation – the process that saw our stem cells give rise to skin cells, bone cells, muscle cells, etc. This radical transformation is all the more impressive when one considers that cells of all types contain the same genome. How do, say, brain cells and blood cells end up with such enormous morphological and functional differences when they contain the same set of genes? Enter epigenetics. As cells differentiate, they accumulate specific epigenetic marks at key positions. These epigenetic marks silence some genes while activating others,[28] resulting in the creation of unique subsets of genes expressed by the different cell types.

Note that when a differentiated cell divides, its descendants do not revert to an undifferentiated state. This is unexpected, because every time DNA is replicated, about half of its epigenetic marks are erased. So how does a new cell "remember" its parental cell type? The surprising answer is that the three-dimensional folding of the genome plays a major role in maintaining cellular identity at a multigenerational scale. Following DNA replication, each cell type's unique genomic spatial structure enables the easy restoration of its signature epigenetic marks. Later, when the cell splits in two, the genome's spatial structure is reorganized. Fortunately, the epigenetic marks are already in place, enabling the speedy restoration of the genome's previous spatial structure. In short, the memory of cellular identity is passed back and forth between epigenetic marks and the genome's three-dimensional shape,[29] with nary a mutation in sight.

Genomic imprinting is another example of a widespread epigenetic phenomenon. All sexually reproducing organisms possess two copies of each gene – one maternal and one paternal. Normally, this would mean that both the maternal and paternal copies of genes are expressed. But in some sexually reproducing organisms, such as mammals, certain genes are only expressed if they are maternal, while others are only expressed if they are paternal. This baffling situation is the result of genomic imprinting, a process in which one copy of a gene is silenced. Genomic imprinting occurs in the zygote, or even earlier, in the sperm and egg cells, or *gametes*, that fuse to produce it. When the zygote and its descendants divide, this parent-of-origin-specific silencing is passed down, resulting in an adult who can only express the maternal/paternal copies of certain genes. Curiously enough, when it comes to generating offspring, all bets are off. The epigenetic marks responsible for genomic imprinting, as it turns out, are summarily erased during gamete production,[30] allowing each zygote to concoct a new genomic imprinting program.

Epigenetic alterations are clearly key modulators of gene expression, but what do they have to do with mutations? Don't they, by definition, operate "above" the playing field of mutations? Yes and no. Although epigenetic marks cannot alter the DNA sequence, they can, and do, promote the occurrence of mutations. Some cancers, for example, develop due to a DNA repair deficiency. But this deficiency is rarely the result of a mutation in a DNA repair gene. In most cases, its root cause is epigenetic marks that reduce the expression of DNA repair genes.[31] Thus, although epigenetic marks do not induce the mutations underlying these cancers, they do prevent their correction, significantly increasing the number of mutations that slip under the radar. How, then, can these mutations be considered random?

In addition to promoting the occurrence of cancer mutations, epigenetic alterations underlie many single-nucleotide mutations, which tend to be distributed in a strikingly nonrandom manner throughout genomes.[32] One kind of epigenetic mark, for instance, can trigger the chemical transformation of a C nucleotide into a T nucleotide.[33] Yes, you read that right – the epigenetic mark doesn't cause a C nucleotide to be swapped out for a T nucleotide, it causes the C nucleotide to *morph into* a T nucleotide.

This bewildering phenomenon may explain why C and G nucleotides are up to six times more mutable than A and T nucleotides,[34] since a C-to-T transition in one strand is always mirrored by a G-to-A transition in the complementary strand. It may also explain why C/G-to-T/A mutations are the most frequent mutations in many bacterial species,[35] and why these mutations make up almost a quarter of all single-nucleotide mutations in humans.[36] But that's not all. The epigenetic mark on a C nucleotide doesn't just increase its likelihood of turning into a T nucleotide; it also appears to trigger an error-prone repair mechanism that results in mutations in neighboring nucleotides.[37] So epigenetic marks not only influence the mutability of "marked" nucleotides, but also have an enormous impact on the mutability of "unmarked" nucleotides in their vicinity. Our *epigenome*, it seems, is the root cause of innumerable nonrandom mutations.

CHAPTER 8

Viral Vagabonds

Mutations, as we've seen, are an inescapable feature of life. But are they equally prevalent in non-life? To answer this almost incongruous question, we must first introduce viruses – non-living entities so bizarre that they are often mistaken for living organisms. Although viruses possess genes, they are not true organisms because they cannot express these genes independently. *Real* organisms have transcription machinery for transferring the information encoded in their genes into RNA, and translation machinery for transferring it from RNA to protein. They also have replication machinery for copying their genes, as well as metabolic machinery for generating the energy required by all of these processes. But viruses have none of these. A typical virus consists of a protein envelope surrounding some DNA or RNA, which constitutes the viral genome. Some viruses also carry around a few proteins, but that's about as far as viral complexity goes.

If viruses lack the machinery that organisms use to copy and express genes, not to mention the machinery for energy generation, how on earth do viruses exist? The plot thickens when one considers that viral genes encode the proteins that make up the viral envelope. So viruses are somehow expressing their genes, and given their ubiquity in nature, they must also be replicating themselves. But how are they pulling this off without the necessary machinery?

The solution to this mystery lies in the way viruses interact with organisms. Despite their non-living status, viruses routinely hijack the cells of living organisms and force them to operate cellular machinery for their own, viral benefit. A virus will typically inject its genome into a cell, and then trick the cell into building viral proteins to be assembled into new viral envelopes. Meanwhile, the cellular replication machinery is duped into replicating the viral genome, generating multiple copies that can be packaged into the viral envelopes to yield a new generation of viruses. Thus, the crafty virus reproduces itself without lifting a finger – an impressive feat for a non-living entity!

Since the viral genome undergoes replication, albeit at the hands of cells, it should come as no surprise that mutations are a staple of viral existence. Some viruses, in fact, are characterized by exceptionally high mutation rates. Take HIV, the deadly virus behind AIDS. Why hasn't an HIV vaccine been developed yet? The answer is that HIV mutates roughly every three replication cycles, resulting in very frequent changes to the meager nine genes in its genome. Consequently, its protein envelope changes so quickly that a vaccine designed to teach the immune system to recognize HIV will be rendered obsolete within just a few replication cycles.[1] The high mutation rate of some viruses is also significant because in viruses even minuscule mutations may have a major impact. Indeed, one study found that a single-nucleotide mutation can turn a certain virus's icosahedral protein envelope into a tetrahedral one about a fifth of the icosahedron's size,[2] highlighting the enormous potential of viral mutations.

So viruses are non-living entities that can nevertheless outmutate living organisms. But are *their* mutations, at least, truly random? To shed light on the nature of viral mutations, one fascinating study examined 15,000 copies of the *SARSCoV-2* genome, unearthing about 6,000 mutations. U nucleotides (the equivalent of our T nucleotides) were found to host a disproportionately high percentage of mutations,[3] suggesting that certain nucleotides are prioritized over others in viral mutagenesis.

This nonrandomness is not just typical of mutations within viral genomes, but is also remarkably common in the mutations viruses cause in *cellular* genomes. But why should viruses cause cellular mutations in the first place? The answer is that some viruses, known as *retroviruses*, have an RNA genome and an unusual modus operandi. Instead of infecting a cell, replicating itself, and going on its merry way, a retrovirus will infect a cell and use its special *reverse transcriptase* enzyme to *reverse transcribe* the viral genome into DNA! The retrovirus then proceeds to integrate the DNA version of its genome into the cellular genome, fusing non-life into life. For the cellular genome, this integration is an insertion mutation, the site of which is of the utmost importance. If the insertion mutation occurs in an insignificant, noncoding region of the genome, it may make no difference whatsoever. On the other hand, if the insertion mutation disrupts an essential gene, it could result in the cell's death.

Retroviral integration sites can also influence disease severity, as in the case of the *Human T-cell Lymphotropic Virus 1*, or *HTLV-1*. This virus causes an inflammatory disease in some of the people it infects, while others remain mysteriously asymptomatic. The key to this disparity lies, surprisingly enough, in the HTLV-1 integration sites. Asymptomatic carriers tend to have integration sites in genes, whereas inflammatory disease sufferers tend to possess integration sites in non-coding regions.[4] The question of how retroviruses "choose" integration sites is therefore a fateful one. Are integration sites "chosen" at random? Or do retroviruses secretly prefer to induce mutations at certain sites?

Some retroviruses, like HIV, demonstrate an obvious predilection for DNA in the peripheral regions of the nucleus[5] as well as for curved DNA.[6] Others, like the *Adeno-associated Virus* (*AAV*), care little for nuclear regions or DNA topology, choosing instead to integrate themselves into a specific chromosome.[7] Retroviral preferences can also be more granular, as in the case of the *Moloney Murine Leukemia Virus* (*MoMLV*), which favors integration sites in active genes,[8] or HTLV-1, which is partial to regions rich in A and T nucleotides.[9] Integration sites are therefore nonrandom with respect to everything from nuclear niches to the nucleotide composition of a sequence, meaning that retroviral insertion mutations are yet another class of nonrandom mutations.

The nonrandomness of retroviral insertion mutations is also evident when organisms benefit from the viral genes inserted into their genomes. Take the curious tale of the *syncytin* protein. In placental mammals like humans, syncytin promotes the cell–cell fusion required for placental development. How did we come to possess such a useful protein? Syncytin was once a viral protein that promoted retrovirus–cell fusion. But following its insertion into mammalian genomes tens of millions of years ago, it was co-opted to form the placenta.

At first glance, this insertion mutation seems charmingly serendipitous. However, a comparison of the genomes of different mammalian lineages reveals that different syncytin genes have embedded themselves into the genomes of various early mammals. Primates, for example, create placentas with the help of syncytins originally encoded by the viral *ERVW1* and *ERVV2* genes, while mice use syncytins originally encoded by the viral *env* gene. As for opossums, their genome contains a degenerated syncytin gene that was once functional, a functional syncytin gene, and a newly acquired syncytin gene that may eventually replace the functional one. The insertion of syncytin genes into mammalian genomes as well as their subsequent pivot to placental development are therefore routine occurrences that cannot possibly be random.[10]

Deep within our genome, a family of retrovirus-like sequences called *transposons* is busy inducing similar insertion mutations. Transposons, you see, are DNA sequences that hop around the genomes of virtually all plants and animals. Unlike a retrovirus, a transposon does not encode envelope proteins to facilitate its transmission from cell to cell. It is, in other words, a grounded retrovirus of sorts, doomed to endless *transpositions* within a single genome. But this limitation does little to detract from its immeasurable evolutionary importance.

Transposons and their remnants make up nearly half of the human genome,[11] and up to 90% of some plant genomes.[12] Transpositions occasionally create new copies of existing genes, which, as we saw in Chapter 2, may subfunctionalize or neofunctionalize to drive evolutionary innovation. Mischievous transpositions can also result in overlapping genes,[13] a surprisingly common phenomenon. And incredibly, transposons were instrumental to the development of antibody-based immunity,[14] which requires constant sequence shuffling to produce a variety of antibodies capable of targeting ever-changing pathogens.

Some transposons jump around the genome by cutting and pasting themselves into new locations. But others, known as *retrotransposons*, stay put and distribute *copies* of themselves far and wide. The secret to this copying and pasting lies in the reverse transcriptase encoded by retrotransposons. That's right – reverse transcriptase, an enzyme previously believed to be strictly (retro)viral, is also encoded by retrotransposons! Thanks to this genetic quirk, a retrotransposon can be transcribed into RNA and then reverse transcribed into DNA, which can be inserted elsewhere in the genome.

But if retrotransposons encode reverse transcriptase, is it possible that other parts of the cellular genome do as well? Cellular reverse transcriptase genes, as it turns out, are not as rare as they may first seem. The human genome, for example, encodes *telomerase*, a specialized reverse transcriptase that elongates the ends of our chromosomes.[15] Similarly, many bacterial genomes encode reverse transcriptases that may play a role in everything from the revival of inactive genes to the creation of new regulatory sequences that modulate gene expression.[16]

Let us return to transposons. The undeniable centrality of transpositions in evolutionary processes forces us to contend with yet another question: are transpositions, and their ensuing insertion mutations, random? Transposition rates definitely aren't, because environmental stress has been shown to increase transposition frequencies in many species. Fruit flies, for instance, respond to extreme heat by boosting transposon-mediated mutagenesis, which accelerates adaptation.[17] But it's not just transposition *rates* that are nonrandom; transposition *sites* must be, too, because transposons are not randomly scattered throughout the genome. Human transposons, for example, often cluster together.[18] Moreover, some mammal transposons make a habit of inserting themselves into the X and Y chromosomes,[19] whereas fruit fly transposons tend to originate in the X chromosome and insert themselves into non-sex chromosomes.[20]

Similarly, some transposons exhibit a preference for insertion into the telomeres at the ends of chromosomes,[21] or into genes.[22] There are also transposons that seek to insert themselves into sequences characterized by very specific nucleotide compositions.[23] In some cases, a potential insertion site is only deemed acceptable if its nucleotide composition resembles the nucleotide composition of the transposon itself, enhancing the tremendous complexity of transposition nonrandomness.[24]

CHAPTER 9

Nano Nomads

Intercellular communication was once thought to be a straightforward affair: one cell releases a molecule, which binds a specific receptor on the surface of another cell, thereby transmitting a signal. But over the last few decades, a growing body of research has elucidated the role played by little particles known as *extracellular vesicles*, or *EVs*, in facilitating intercellular communication. Like cells, EVs are enveloped by membranes, although they are incapable of division. EVs are released by a wide range of cell types, and are commonplace in everything from bacteria to plants. Their content – DNA, RNA, proteins, organelles, etc. – varies greatly, and is derived from their cells of origin. Thus, when an EV is taken up by a cell, the composition of biological molecules within it constitutes a specific signal which may alter the recipient cell's behavior.

There are many known types of EVs, and probably quite a few types that have yet to be discovered. But for now, we will expatiate on five EV types to gain a fuller understanding of how these intriguing particles work. Let us start with *ectosomes*, the RNA- and protein-containing EVs that bud directly from the cell membrane into the extracellular environment. Ectosomes are regularly released at a leisurely rate by resting cells, though cell stimulation can induce the release of an ectosome barrage. Many cells release ectosomes directly into the bloodstream, allowing them to travel great distances before reaching faraway target cells. Having reached their target cells, ectosomes fuse with these cells' membranes to deliver their cargo, often eliciting a change in the target cells' gene expression. Thus, ectosomes play a key role in the intercellular communication underlying processes such as inflammation and tumor metastasis.[1]

Unlike ectosomes, *exosomes* are EVs that bud from the cell membrane *into* the cell before eventually being released from it. In addition to DNA and RNA,[2] exosomes may contain up to 20,000 proteins, some of which are present in nearly all exosomes.[3] And although exosomes are routinely released from healthy cells, their release from tumor cells is of special interest due to their ability to orchestrate the metastatic process. Tumor cells in the bloodstream can, and do, reach almost every organ, which is why the existence of predictable metastasis sites is so bewildering. Why, for example, does melanoma usually metastasize to the lungs, while colon cancer usually metastasizes to the liver? Shouldn't tumor cells establish metastases wherever they happen to land? This seemingly inexplicable phenomenon makes perfect sense when exosomes are factored in. Long before tumors release *cells* into the bloodstream, they release special *exosomes*, which wander around the body and set up shop in specific organs. Thus, these organs are primed to receive and nurture tumor cells,[4] much like tilled soil is primed to receive and nurture seeds.

Apoptotic bodies are yet another type of EV. Their unusual name is derived from *apoptosis*, a kind of cellular suicide. Why, you may be wondering, would cells commit suicide? The answer is that there are two ways cells can die. The first is by bursting and allowing their (often toxic) contents to spread – a surefire method for damaging, and potentially killing, neighboring cells. The second and altogether friendlier way cells can die is by self-degrading and collapsing quietly into themselves, thus eliminating collateral damage. This is known as apoptosis, and it is a vital biological process. During embryonic development, for instance, the apoptosis of cells in certain anatomical locations is a necessary step in organ development. And when cells acquire cancerous mutations, they often "choose" to undergo apoptosis for the greater good.

But what does apoptosis have to do with EVs? As it turns out, when apoptotic cells collapse into themselves, their contents are divided into little membrane-enclosed vesicles. These apoptotic bodies, as they are commonly known, can consequently contain everything from RNA[5] to intact organelles,[6] and are typically engulfed by certain kinds of immune cells.[7] Moreover, if apoptotic bodies are *not* cleared away properly, their persistence may contribute to the development of autoimmune diseases.[8]

Migrasomes, an equally fascinating type of EV, can contain everything from RNA to damaged organelles destined for disposal.[9] They appear when a cell migrates by extending finger-like protrusions in the direction it wishes to go and allowing these protrusions to attach to surrounding surfaces. As the migrating cell inches along, the protrusions at its rear break to detach from their surfaces, leaving migrasomes in their wake.[10] The formation of migrasomes may seem, at first glance, almost accidental, and therefore most likely inconsequential. But migrasomes (or their contents) are often gobbled up by migrating cells as they pass by the area previously inhabited by the migrasomes' cells of origin. This, of course, renders migrasomes ideal candidates for the transmission of spatial and biological information in dynamic cellular systems.[11]

So do migrasomes help coordinate processes involving multiple migrating cells? Absolutely. We know, for instance, that they are involved in regulating organogenesis in embryos, because their absence can lead to the development of defective organs.[12] They are also frequently released from the platelets of COVID-19 patients, for reasons currently unknown.[13]

Our final type of EV is an unexpected one, as it consists of a family of *viruses*. This may sound strange at first, but some viruses have a membrane around their protein envelope which they acquire while budding out of the host cell's membrane, just as EVs do. These membrane-bound viruses can therefore be classified as another type of EV, one that is released by cells under viral attack. Furthermore, the viruses themselves are not the only EVs released under these circumstances. When the

Hepatitis C Virus (*HCV*) infects cells, for example, a massive release of defective viruses ensues. These noninfectious viruses are often called *subviral particles* because they contain viral proteins and very little else. But if subviral particles are noninfectious, why are they released in the first place?

One possible answer is that everything from the formation of these peculiar EVs to their release is purely accidental. Given the chaos reigning inside a cell hijacked by a virus, it stands to reason that as the cell stuffs viral genomes into a fleet of viral envelopes, it will make the occasional blunder, resulting in a defective particle. This explanation, though plausible-sounding, falls apart when one considers that the concentration of subviral particles in the blood of HCV patients can be up to 10,000 times higher than the concentration of infectious viruses. This is clearly not accidental, but why are subviral particles so prevalent?

One beguiling possibility is that HCV intentionally induces the release of copious numbers of subviral particles in order to overwhelm the immune system and distract it from the real, infectious viruses.[14] HCV, in other words, may be a criminal mastermind that causes cells to release certain EVs, known as subviral particles, as an immune evasion strategy to protect other EVs, known as infectious viruses.

Membrane-bound viruses are, as we've seen, undoubtedly EVs. But perhaps this rigid classification is the wrong way to approach the marvelous world of viruses and their derivatives. A better way to understand them may be to conceive of a continuum of membranous particles released by cells, with "classic" EVs at one end and membrane-bound viruses at the other. The same idea, incidentally, may be applied to the viruses discussed in Chapter 8. A mobile gene continuum can provide conceptual space for retroviruses and transposons at either end, just as a membranous particle continuum can help us make sense of subcellular reality. Importantly, this approach can help us formulate profound questions, such as: which came first, the EV or the virus? Did viruses hijack preexisting EV-budding pathways, or did cells repurpose viral budding mechanisms to release EVs? There is, of course, also a third option: EV release and viral release may have arisen simultaneously before acquiring shared evolutionary tricks.

Having sampled the richness of the EV world, let us zoom in on one integral component of EVs – RNA. Perceptive readers may have noticed that every one of the EV types discussed earlier contains RNA, a tidbit that may appear insignificant at first. RNA, however, is more than just another molecule that EVs lug around; it is the key to a revolutionary method of intercellular communication. This was demonstrated in a groundbreaking study that exposed human cells to EVs released from mouse cells. When the mouse EVs were taken up by human cells, the RNA in the EVs was promptly translated by the human cells, yielding functional mouse proteins![15] This is a stunning discovery, because a cell's proteins were once thought to be the products of its own genes. However, it seems that when EVs shuttle RNA from cell to cell, they routinely induce the expression of *alien* proteins, challenging our preconceived notions of discrete cellhood.

But what do EVs and their RNA content have to do with mutations? Well, one could argue that the introduction of RNA into cells by EVs constitutes a novel kind of insertion mutation. Readers may object to this argument because mutations are usually defined as changes to *DNA*, but this is an arbitrary definition at best. After all, some viruses, as we saw in Chapter 8, possess RNA genomes, and changes to these genomes are called mutations even though no DNA is involved. Consequently, expanding the definition of mutations to include any changes to *protein-encoding material* is the only logical way forward. Armed with this expanded definition, we can classify the EV-borne RNA genes that burst into cells to produce strange proteins as insertion mutations. This, you will find, is a much clearer lens through which to view life, because our real genome – the one made up of both DNA *and* RNA – is in a constant state of flux.

Having established that EVs can induce mutations, we now turn to the question of the randomness of these mutations. EVs may at first seem to be perfectly suited to random mutation induction. They are, after all, released "into the wild", where they should theoretically be taken up by a completely random selection of cells. But the truth is that EVs aren't randomly dispersed; they are purposefully dispatched to specific target cells. This targeting is designed to achieve various objectives, such as warning distant cells of a dangerous pathogen.[16] Cells may also send EVs to specific cells with which they wish to cooperate,[17] or target cells they wish to harm. A good example of the latter is the EV-mediated warfare employed by the *A. thaliana* plant introduced in Chapter 6. When this cress is infected by the *Botrytis cinerea* fungus, it releases EVs that are taken up by fungal cells. The fungus then translates EV-borne RNA, producing sinister proteins that compromise its infection of the plant.[18] A cell, then, may utilize EVs to induce everything from friendly warning mutations to hostile mutations in other cells. Needless to say, nothing about this sensational mechanism is even remotely random.

CHAPTER 10

Invisible Instigators

Radioactivity is perhaps the most infamous of mutagens. From the nuclear bombings of Hiroshima and Nagasaki to the nuclear disasters at Chernobyl and Fukushima, the catastrophic effects of radiation have captured the public's imagination. However, when this doomsday force is unleashed, it does not result in the post-apocalyptic wastelands or mutant ninja turtles that consumers of pop culture have come to expect. In fact, after the Chernobyl disaster, (largely normal) life continued in the surrounding area. How is this possible? The answer is that when radiation damages DNA and induces its mutation, this leads to three possible outcomes. Many mutations, as we saw in Chapter 3, are entirely inconsequential, while others are so deleterious that they quickly kill their carriers. This leaves us with the mutations of intermediate severity – the ones with tangible, but nonlethal effects.

Thus, the mutations most commonly observed in the aftermath of nuclear incidents are those that alter organ size, shape, number, or color. Following the bombing of Hiroshima, for example, an unusually high percentage of survivors went on to have children with small heads.[1] Similarly, the Fukushima accident yielded butterflies with significantly smaller wings.[2] The trees in Chernobyl, on the other hand, began to produce giant, odd-shaped leaves in unprecedented hues after being exposed to radiation.[3]

The mutations behind these peculiar traits may seem to be random, but research suggests otherwise. Radiation often causes double-stranded breaks in DNA, and the sites of these breaks are nonrandomly distributed in the genome.[4] Radiation is also extremely likely to cause deletion mutations, which tend to cluster in certain genomic areas.[5] Speaking of deletion mutations, a study on wheat grown in the Chernobyl area revealed that the deletion of a specific kind of sequence was favored over others. The study, which compared wheat plants grown for a single generation in Chernobyl with unexposed control plants, demonstrated a radioactive bias towards the deletion or insertion of *repetitive* sequences. It also uncovered a high frequency of insertions of sequences of unknown origin,[6] implying that we have much to learn about radioactive mutagenesis.

Mutations induced by radiation are, surprisingly enough, also nonrandom in more "conventional" ways. This is evident in the Kerala region of India, which boasts the world's highest levels of natural radioactivity in a densely populated area. To elucidate the mutational effects of lifelong exposure to natural radioactivity, researchers sequenced the mitochondrial DNA of Keralese individuals and compared it to the mitochondrial DNA of inhabitants of the neighboring, low-radiation islands. Previous studies had failed to detect in the Keralese an increased risk for the DNA breaks and large-scale mutations that are so typical of organisms exposed to radioactivity. This study, however, revealed that contrary to popular belief, radiation can accelerate single-nucleotide mutations. The Keralese single-nucleotide mutation rate, in fact, was twelve times higher than the control group's rate – but only in mutation hotspots.[7] The nonrandomness of radioactive mutations, in other words, isn't necessarily outlandish; it can also be a somewhat exaggerated version of "normal" mutational nonrandomness.

Exposure to radiation can, of course, result in one of the phenomena most commonly associated with mutations – the development of cancer. But why, and *how*, do mutations play such a significant role in this terrible disease? To shed light on this fascinating subject, we must first delve into the two main types of genes whose mutations are implicated in cancerogenesis.

The first of these are the *oncogenes*, which start out as *proto-oncogenes*, or genes responsible for cell survival, growth, or division. Proto-oncogenes are normally vital to the cell's development, but if they mutate, they may come to be permanently activated, at which point they become oncogenes. And oncogenes are dangerous because too much of a good thing can prove to be a very bad thing. A proto-oncogene tasked with, say, promoting cell division is undeniably a good thing, because it enables zygotes to develop into adult bodies, wounds to heal, and tissues to renew themselves. However, if this proto-oncogene mutates into an oncogene that is *always* promoting cell division, regardless of the body's needs, out-of-control cell division will soon give rise to a mass of cells known as a *tumor*. A good example of an oncogene is the *Philadelphia chromosome*, a mutant fusion of segments of human chromosomes 9 and 22. This fusion transforms a proto-oncogene involved in cell division into an oncogene that promotes uncontrollable cell division, resulting in a kind of leukemia.[8]

Tumor suppressor genes are the second type of gene implicated in cancer. These genes normally slow down cell division or nudge cells to undergo apoptosis if they sustain irreversible damage, such as the acquisition of an oncogene. (Apoptosis, you may recall, is a kind of programmed cellular suicide that is usually triggered when cells are irreversibly damaged.) Consequently, if mutations interfere with the proper functioning of tumor suppressor genes, cells are free to divide like crazy – a surefire way to create a tumor. Some tumor suppressor genes are *DNA repair genes*, although these are often seen as a separate type of cancerous gene. DNA repair genes normally fix mutations incurred during DNA replication, but if they find themselves unable to do so, they trigger apoptosis. It should therefore come as no surprise that when a mutation in a DNA repair gene renders it dysfunctional, mutations in other genes go uncorrected, and that despite this dangerous state of affairs, the cell does not undergo apoptosis. Thus, mutations in DNA repair genes may facilitate the establishment of, say, mutations in proto-oncogenes, paving the way for tumor development.

For examples of DNA repair genes, look no further than the famous BRCA1 and BRCA2, two unrelated genes associated with an increased risk of breast and ovarian cancer. BRCA1 and BRCA2 play an important role in repairing DNA damage, especially when this damage consists of double-strand breaks. But if either of these genes is damaged by a BRCA mutation, double-strand breaks can no longer be repaired properly, which is why cancer becomes so likely.[9]

Cancerous tumors, it should be noted, are the result of multiple mutations in both proto-oncogenes *and* tumor suppressor genes. But how many mutations does it take to make a tumor? When it comes to tumor suppressor genes, two mutations are a prerequisite for cancer. This is known as the *two-hit hypothesis*, and it is based on a genetic redundancy of sorts. Every individual, you see, possesses two copies of each tumor suppressor gene – one inherited from the mother and one inherited from the father. Consequently, if one copy is inactivated by a mutation, the second copy is still there to compensate for it. But if the second copy is "hit", too, then no functional copies of the gene remain and nothing stands in the way of tumor development.[10]

Certain kinds of mutations, as we've seen, are responsible for jump-starting cancerogenesis. But when cells become cancerous, they immediately start to accumulate additional mutations, and these only make things worse. Some of these mutations allow them to continue dividing indefinitely. This is actually a remarkable feat, because normal cells can only divide a limited number of times due to the shortening of their telomeres. The telomeres, you may recall, are regions of repetitive sequences found at the ends of chromosomes. Before a cell divides, it must replicate its chromosomes, but every time a chromosome is replicated, its telomeres are shortened. Eventually, the telomeres are whittled down so far that any additional shortening will gnaw at the genes nestled in the chromosome's interior. To prevent this undesirable outcome, a normal cell becomes *senescent* and stops dividing. Cancerous cells bypass this barrier by rapidly acquiring mutations that activate the *telomerase* gene. This gene encodes an enzyme that elongates telomeres during embryonic development, allowing cells to divide over and over again to create our organs. Once our organs are fully formed, the telomerase gene is permanently inactivated – or at least it should be. Mutations in cancerous cells tend to reactivate it, bestowing limitless division capabilities on these unruly cells.

Having achieved limitless division potential, the rapidly growing numbers of cancerous cells are soon faced with the problem of nutrient scarcity. When a tumor first develops in a tissue, it leeches off the local blood supply, but if it keeps growing, the amount of blood required to sustain its many cells soon surpasses the amount the circulatory system delivers to the host tissue. This is where *angiogenic mutations* come in. These intriguing mutations cause new blood vessels to be formed in the tumor's vicinity, so that the tumor may enjoy an increased flow of oxygen-rich and nutrient-rich blood. The tumor is now free to grow at an unprecedented rate, which it does with the help of three additional types of cancerous mutations.

The first type consists of mutations that provide *self-sufficiency in growth signals.* The sheer number of cells that make up our bodies and the complexity of the organs they form clearly require meticulous intercellular coordination. It is therefore unsurprising that cells normally divide in response to growth signals from other cells, which instruct them to undergo divisions at specific times. But cancerous cells quickly lose this dependency on external signals and may even start producing their own growth signals, which promote the excessive divisions that form tumors.

Mutations that provide *insensitivity to anti-growth signals* are crucial to cancerogenesis for a similar reason. Normal cells, you see, heed anti-growth signals from other cells, which instruct them to stop dividing once a tissue has been formed or a wound has been healed. But cancerous cells acquire mutations that allow them to ignore these signals and carry on dividing indefinitely.

Taken together, the mutations that provide self-sufficiency in growth signals and the mutations that provide insensitivity to anti-growth signals remove all barriers to unbridled cell division, though they do nothing to solve the obstinate problem of apoptosis. So what good is dividing tirelessly to yield a prodigious number of defective cells destined for apoptosis?

This brings us to the final type of mutation that facilitates tumor growth. *Apoptosis evasion mutations*, as they are known, allow even the most abnormal cancerous cells to escape apoptosis, paving the way for truly uninhibited tumor growth. Growth, however, is not the be-all and end-all of tumors. Once mutations ensure a tumor's unrestrained growth, other mutations may allow cells to break off the tumor, invade neighboring tissues, and *metastasize*, or spread to distant organs.[11] As these migrating cancerous cells invade blood vessels and make their way around the body, they are more vulnerable than ever to immune system attacks. Consequently, they tend to acquire immune system evasion mutations that allow them to escape detection.[12]

Cancerous cells, as we've seen, acquire a great number of mutations which remove various barriers to cancer progression. However, not all the mutations in cancerous cells help them achieve their sinister goals. Indeed, the copious amounts of mutations bombarding a cancerous cell's genome can be divided into *driver mutations* and *passenger mutations*. Driver mutations help cancerous cells survive or divide in a hostile environment – these are the mutations we have been discussing thus far. Passenger mutations, on the other hand, provide no benefit to cancerous cells, and are just "hitchhikers" taking advantage of the accelerated mutagenesis within these cells. Just how often do these hitchhikers appear in cancer? It is estimated that roughly 97% of the mutations in cancerous cells are passenger mutations,[13] meaning that driver mutations are actually relatively rare.

Cancerous mutations are a maliciously diverse bunch, but are they random? Before we answer this question, it should be noted that cancerous mutations are innately nonrandom, because mutations in genes such as DNA repair genes facilitate the occurrence of more mutations, leading to a mutation spree that would never be tolerated in a normal cell. Put simply, since cancerous cells are characterized by abnormally high mutation rates,[14] it is inherently absurd to assume that their mutations are random. Moreover, if mutations are only considered random if they are equally likely to arise at any age, then cancerous mutations, which predominantly affect older individuals, are decidedly nonrandom. The reason for this apparent ageism is simple: DNA repair mechanisms become less efficient with age,[15] resulting in a higher incidence of both cancerous and noncancerous mutations.

Setting aside the likelihood and timing of their appearance, are cancerous mutations random? Not at all. In humans, cancerous mutations cluster instead of occurring randomly throughout the genome.[16] Probing the mutations of human cancerous cells also reveals that regions of closed chromatin, where, as you may recall, the DNA is densely packaged, are characterized by a higher mutation density than regions of the loosely packaged open chromatin.[17] This interesting disparity is probably due to the DNA repair proteins' difficulty in accessing closed chromatin,[18] which results in a higher percentage of uncorrected mutations. When it comes to single amino acid substitutions, driver mutations display a preference for genomic regions that are *evolutionarily conserved*, which means that they have changed little throughout evolutionary history and are therefore shared by many species. Single amino acid substitution passenger mutations, on the other hand, are less likely to occur in evolutionarily conserved regions.[19]

Chromosomal cancerous mutations, as it turns out, are also nonrandom, as some chromosomes are more likely to be affected than others. Interestingly, even the chromosome breakage events that precede chromosomal cancerous mutations are clearly nonrandom. These breakage events consistently occur in *common fragile sites*, or *CFSs*, which tend to consist of sequences predisposed to folding into unusual three-dimensional DNA structures. These unusual structures, in turn, tend to hinder DNA replication, resulting in the breakage of newly synthesized chromosomes at CFSs. But that's not all – in addition to the CFSs' replication woes, which often lead to breakage events in pre-cancerous cells, CFSs are a popular destination for viruses looking to integrate themselves into the genome.[20] Needless to say, viral integrations can cause everything from cervical cancer to leukemia,[21] implicating CFSs in yet another cancer development mechanism.

If cancerous mutations were truly random, we would see cancers arise as the result of chromosome breakage or viral integration at random positions in the genome. The disproportionately high number of breakage events and viral integrations in CFSs clearly demonstrates that the mutations responsible for kickstarting cancer are nonrandom, but are later cancerous mutations, at least, random? Of course not. Chromosome deletion mutations are exceedingly common in cancer,[22] and many of these deletions include CFSs. Moreover, many large genes found in CFSs are inactivated in a variety of cancers,[23] indicating that CFS mutations play an openly nonrandom role in cancer progression.

CHAPTER 11

Targeted Tethering

Fusion mutations, or mutations that fuse two genes to form a new one, are a defining feature of many cancers. In fact, fusion mutations are strongly suspected of being major driver mutations in multiple cancer types,[1] which is why understanding the mechanisms that fuse genes is so critical. Some fusion mutations are actually deletion mutations, because when a segment of DNA between two genes is deleted, the two genes will end up fused together. Other fusions are the products of duplication mutations that generate multiple, adjacent copies of a gene. And a third group of fusion mutations occurs when genes are *translocated* to new locations – either within the same chromosome or on other chromosomes.[2]

These translocations, incidentally, may be caused by *chromothripsis*, a catastrophic event of chromosomal shattering followed by a messy repair process that patches chromosomal fragments together in no particular order or orientation. Although chromothripsis may sound outlandish, it is a hallmark of tumor development, which, as we saw in Chapter 10, induces myriad mutations. Chromothripsis has also been shown to occur in cases of apoptosis, the cellular suicide we encountered in earlier chapters. Normally, when a cell undergoes apoptosis, its DNA is irreversibly fragmented. But every once in a while, an apoptosizing cell manages to claw its way back to life by hastily repairing its genome. Repairing a fragmented genome correctly is, of course, a colossal undertaking that the cell could scarcely hope to achieve under the best of circumstances. It should therefore come as no surprise that attempting to do so in the course of abortive apoptosis is virtually guaranteed to result in chromothripsis,[3] which, in turn, results in numerous fusion mutations.

Fusion mutations have been shown to play a key role in protein evolution. As we saw in Chapter 3, proteins contain segments known as domains which serve specific purposes, such as binding other molecules, cleaving molecules, or joining molecules. But how do radically different domains come to be linked together in *multidomain proteins*? In other words, how do DNA sequences encoding a variety of protein domains come to inhabit the same gene? This is where gene fusion comes in.[4] Although "recent" fusion mutations produce clumsy chimeric genes, it is easy to see how a series of such fusions occurring in the very distant past could piece modern genes together, domain by domain. This evolutionary process is not just hypothetical – a study on eighteen species of bacteria demonstrated that up to 1% of the genes in a bacterial genome are the product of fusion mutations.[5]

So how do these alluring fusions transpire? And are they random? To get to the bottom of this mystery, let us start by examining two seemingly unrelated primate proteins. The first of these, *Cyclophilin A*, or *CypA* for short, can bind several retroviruses. The second protein, *Tripartite Motif Containing 5*, or *TRIM5* for short, detects retroviruses and blocks their replication. You can probably see where I'm going with this: long ago, a fusion mutation between the genes encoding CypA and TRIM5 produced a novel protein which provides resistance to certain retroviruses.[6] This fusion mutation, however, couldn't have been purely accidental, because it occurred at least twice – in different primate species![7] But that's not all – TRIM5 belongs to the TRIM protein family, which means that there are other TRIM genes with which the CypA gene could potentially fuse. Now, if fusion mutations arise when genes randomly bump into each other, we should expect to see a variety of CypA–TRIM fusion proteins. In practice, CypA has only fused with TRIM5, but why is this the case? And would fusions with other TRIM genes be similarly beneficial?

When researchers sought to answer these questions by engineering fusion mutations between CypA and various members of the TRIM family, the resulting fusion proteins provided *less* resistance to retroviruses than the fusion between CypA and TRIM5.[8] Let that sink in – despite there being multiple TRIM genes, at least two copies of the CypA gene have independently fused with the TRIM5 gene, yielding the best possible outcome. It is almost as though CypA has repeatedly singled out TRIM5 for fusion, producing a protein that maximizes antiviral protection.

How can the "right" TRIM gene be singled out for fusion? The nonrandomness of this specificity, though perplexing, is just the tip of the iceberg. Comparisons of the genomes of closely related species reveal that some pairs of genes fuse in one species but remain separate in others. And incredibly, it is the genes that interact in one species that tend to fuse elsewhere. Moreover, the more tightly these genes interact as separate entities, the more likely they are to fuse together in other species, suggesting that fusion mutations favor genes that are already functionally intertwined.

Proponents of mutational randomness would be hard pressed to explain this phenomenon. They could, perhaps, posit that we have it all backwards – fusion mutations force genes to interact, and if a mutation later *unfuses* two genes, they will continue to interact as distinct genes.[9] In other words, an argument could be made that some of the genes that have been shown to interact in one species and to be fused in another were actually fused in the ancestor of both species before splitting apart in one lineage. This is a compelling argument, but does it hold water?

Before answering this question, it should be noted that recurrent fusion mutations didn't just pop up throughout our evolutionary history; they also feature prominently in various human cancers. And remarkably, the genes that fuse repeatedly in human cancers are often the same genes that have fused over the course of other primates' evolution. Now, if these genes were fused and later unfused in our ancestors, why would they *re-fuse* in our tumors? This makes little sense when one compares the selective forces acting on individuals with the selective forces acting on tumors. Natural selection favors mutations that increase an individual's ability to survive and reproduce, which means that if the fusion of two genes persists over many generations, it is probably conducive to survival and/or reproduction. Selection in tumors, on the other hand, favors the rapid division of cancerous cells – which is detrimental to the individual's survival and reproduction. We should therefore expect there to be little overlap between genes that have fused throughout primate evolution and genes that fuse over and over again in human cancers. But this overlap does exist, providing unequivocal proof that interacting genes are more likely to undergo fusion mutations, regardless of their cellular circumstances.[10]

The patient reader will, I hope, allow me one final layer of nonrandomness. Fusion mutations have been shown to be influenced by genomic distance, because a gene is more likely to fuse with a neighboring gene than with a distant one. Does this finding contradict the evidence that fusion mutations are largely the result of gene–gene interactions? Not at all. It appears that when distant genes interact, they are likely to undergo translocation mutations that bring them closer together, setting the stage for an eventual fusion mutation.[11]

CHAPTER 12

Reproductive Revolutions

Mutations are at the heart of one of the most bewildering biological phenomena: sexual reproduction. Why do many species "choose" to reproduce sexually, given that asexual reproduction is a much simpler propagation method practiced by species ranging from bacteria to some types of sharks?[1] The genetic variation gained by mixing genes at the onset of every generation is widely believed to be the purpose of sexual reproduction, but this is only part of the story. After all, there are easier ways to mix genes, such as the horizontal gene transfer discussed in Chapter 4. So what is sexual reproduction *really* for?

One could argue that the main purpose of sexual reproduction is to eliminate deleterious mutations. Think about it this way: organisms are constantly acquiring slightly deleterious mutations – the kind that aren't bad enough to kill them, but that are nonetheless disadvantageous. For an asexual organism, the acquisition of such a mutation is virtually irreversible. An additional mutation could, hypothetically, reverse the deleterious mutation, but in most cases the affected asexual organism and all its descendants will be doomed to carry the deleterious mutation for all eternity. Sexual organisms, on the other hand, only pass down around half of their genes to each offspring, and are therefore capable of eliminating a deleterious mutation from their lineage in as little as a single generation. Viewed this way, sexual reproduction isn't just about genetic variation; it's mostly about giving every generation a chance at a fresh start.

This is an elegant hypothesis, but is there any evidence to support it? Absolutely! Sexual organisms have been shown to accumulate beneficial mutations at a much higher rate and deleterious mutations at a much lower rate than their asexual counterparts. And interestingly, self-fertilizing organisms, such as some plants, are intermediates between sexual and asexual organisms.[2]

Sexual reproduction, as we've seen, appears to have evolved to allow organisms to discard undesirable mutations. But once this fascinating process arose, it began to yield all sorts of novel mutations. Take, for example, the mutations responsible for determining the male-to-female ratio. Sexual reproduction requires the existence of at least two sexes, but how many of each should a species produce? As humans, we tend to take a 1:1 sex ratio for granted, by why did this egalitarian ratio arise in the first place?

The answer becomes obvious when one conducts a little thought experiment. Imagine that one male was born for, say, every two females. Males, in this scenario, would enjoy an advantage over females, because females would be fighting over every available male, and not the other way around. Consequently, many males would copulate with multiple females, and produce numerous offspring, while most females would produce considerably fewer offspring due to the scarcity of males. Under these circumstances, any mutation that increased an individual's chances of producing male offspring would be enormously advantageous. But should such a mutation arise, it would quickly lead to an abundance of male offspring, which would, paradoxically, render males more common than females. This, of course, would tip the scale in favor of females, who would suddenly have their pick of males. Consequently, any mutation that increased an individual's chances of producing *female* offspring would now become advantageous, causing females to outnumber males again. This endless oscillation can be avoided by just sticking to the 1:1 sex ratio we know and love, but is this ratio set in stone?

At first glance, sex determination in wood lemmings seems identical to sex determination in humans, since females have two X chromosomes and males have one X and one Y chromosome. But a closer look reveals the startling existence of normal, fertile females with an X chromosome and a Y chromosome – females that "should" be males! This bizarre phenomena is the result of a *sex-reversal mutation*. The X chromosome of these peculiar females, you see, contains a specific deletion mutation that suppresses the Y chromosome's male-determining activity.[3] Thus, a mutation manages to skew the 1:1 sex ratio *already established* by chromosomal composition.

But that's not all – mutations can also alter the sex ratio *before* conception. In humans, a single-nucleotide substitution with the catchy name of *rs144724107* has been shown to increase the probability of having a girl by a remarkable 10%. Little is known about how this mutation influences the sex ratio, but since it occurs in the vicinity of the *ADAMTS14* gene, which is involved in fertilization,[4] we can safely assume that it tips the scale in favor of females very early on.

Sexual reproduction also opens the door to some highly unconventional mutations, such as the chromosome deletion mutation that renders yeast bisexual,[5] or the even zanier insertion mutations that may be induced by *sperm-mediated gene transfer*. This form of gene transfer, which has been hypothesized, sounds complicated, but it is actually surprisingly simple. The female reproductive tract contains "free" DNA molecules derived from dead cells. The uptake of these DNA molecules, or *exogenes*, by sperm cells navigating the female tract is highly likely, as is the possibility of their incorporation into the egg cell at fertilization. This, of course, is a kind of insertion mutation, because exogenes aren't "natural" components of the zygote's genome. Still, they may end up integrated into our genomes,[6] forcing us to reconsider everything we thought we knew about our genetic origins.

In addition to enabling the occurrence of unorthodox mutations, sexual reproduction promotes markedly nonrandom mutations in several ways. Mutations that give rise to new genes, for instance, exhibit an undeniable preference for sperm cells,[7] instead of popping up in random cells throughout the body. Similarly, the vast majority of human mutations occur in males, instead of being evenly distributed between the sexes. The reason for this disparity is deceptively simple: unlike egg cells, which stop dividing during embryonic development, the germ cells that give rise to sperm cells divide about every two weeks from puberty onwards. In practice, this means that by the time a human male turns thirty, the number of divisions undergone by his germ cells is a whopping fifteen times higher than the number of germ cell divisions undergone in a thirty-year-old female.[8] Consequently, males have countless more opportunities to acquire mutations, a fact reflected in the sheer number of mutations males introduce into the human gene pool. But that's not all – older fathers pass on more mutations than younger fathers, simply because they have had more time to accumulate mutations in their ever-dividing germ cells. Thus, fathers pass on an additional two mutations or so every year after puberty, unlike mothers, who pass on a largely unchanging number of mutations to their offspring.[9]

A close examination of the human X and Y chromosomes, or *sex chromosomes*, uncovers yet another trove of nonrandom mutations. The X and Y chromosomes, it should be noted, evolved from an identical chromosome pair that existed roughly 180 million years ago, and were therefore presumably once subject to identical mutational processes. But long ago, large portions of the proto-Y chromosome appear to have undergone inversion mutations that sparked its divergence from the proto-X chromosome.

This divergence, however, wasn't just a matter of acquiring different mutations; the two chromosomes embarked on radically different mutational journeys. The proto-Y chromosome went on to acquire so many deletion mutations that it lost a jaw-dropping 97% of its genes. Meanwhile, the proto-X chromosome managed to hang on to both its genes *and* their sequential order,[10] an achievement made possible by its unusually low mutation rate.[11] The Y chromosome's high mutation rate[12] and its predilection for deletion mutations may eventually lead to its disappearance,[13] which could necessitate the recruitment of new sex chromosomes.

Although the transformation of non-sex chromosomes, or *autosomes*, into sex chromosomes may appear to be a singular evolutionary event, it is actually an exceedingly common one. The degeneration of Y chromosomes throughout evolutionary history has repeatedly driven various fish, amphibian, and reptile species to convert autosomes to sex chromosomes via a gradual mutational process. And every time a "new" Y chromosome degenerated, autosomes were yet again transformed into sex chromosomes in an endless cycle of turnovers. These astounding turnovers, however, do not occur at random. Autosomes that were once sex chromosomes are very likely to use their "time off" to purge themselves of deleterious mutations before being re-selected to become sex chromosomes.[14]

This extraordinary nonrandomness was demonstrated in a wonderful study on twenty-eight true frog species which had collectively undergone at least thirteen sex chromosome turnovers. True frogs have thirteen chromosome pairs, which means that if autosomes were randomly selected for turnovers, we would expect each chromosome pair to serve as sex chromosomes roughly once in true frog evolutionary history. But in practice, the same five chromosome pairs were always tapped, and three of these five pairs were shown to be especially popular sex chromosome candidates.[15] Sexual reproduction, then, seems to have started out as a mechanism for discarding deleterious mutations, but somewhere along the way, it morphed into a nonrandom mutation generator.

CHAPTER 13

Dizzying Duplications

The most influential kind of mutation, evolutionarily speaking, is probably the *whole-genome duplication*, or *WGD*. Mutations, as we've seen, can duplicate single genes or even multi-gene chunks of chromosomes, but every so often, a WGD will double a cell's chromosome number, resulting in *polyploidy* – the possession of "excessive" sets of chromosomes. To understand how this extraordinary mutation occurs, we must first understand how cells normally come to possess the "right" number of chromosomes, and why this "right" number varies among cell types and even species.

Let us start with humans. Most of the cells that make up the human body are somatic cells that contain two sets of chromosomes – one inherited from the father and one inherited from the mother. Before dividing, a somatic cell dutifully replicates these two sets of chromosomes, producing four sets of chromosomes that are subsequently divvied up between the two new somatic cells. But when cells in the ovaries and testicles divide to produce gametes, a different kind of division takes place. Instead of replicating its chromosomes once and dividing once, the ancestor of gametes will replicate its chromosomes once and then undergo *two* rounds of divisions, yielding four cells. Consequently, each of the four cells will contain just one set of chromosomes – exactly half of what a somatic cell contains.

This is how gametes are produced. But what if something goes wrong and a gamete ends up with two sets of chromosomes? If this *unreduced* gamete, as it is known, later fuses with a "normal" gamete during fertilization, the resulting zygote will end up with three sets of chromosomes. WGDs can also occur if somatic cells undergo abnormal cell division or if an egg cell is fertilized by multiple sperm cells, and hence they are much more probable than they first seem.

These monumental mutations are responsible for many kinds of polyploids, such as *triploids* (with three sets of chromosomes), *tetraploids* (with four sets of chromosomes), *pentaploids* (with five sets of chromosomes), *hexaploids* (with six sets of chromosomes), *heptaploids* or *septaploids* (with seven sets of chromosomes), *octaploids* (with eight sets of chromosomes), etc. There are even *tetratetracontaploids*, with an astonishing forty-four sets of chromosomes![1] Moreover, closely related organisms can be different kinds of polyploids. Durum wheat, for example, is a tetraploid strain, whereas bread wheat is a hexaploid one.[2]

In a way, we are all polyploids, since all organisms are *paleopolyploids*, or descendants of ancient polyploids.[3] This is not immediately obvious, because WGDs that occurred in the very distant past may go undetected due to the diverging mutational trajectories of the "extra" chromosome sets, which often render them unrecognizable.[4] The ancestor of all vertebrates, for instance, underwent two WGDs, the signs of which were gradually effaced from our apparently non-polyploid genome.[5]

Does the universality of paleopolyploidy mean that WGDs must be beneficial mutations? Not at all. WGDs, as it turns out, are largely detrimental due to the replication and division difficulties they create. Consequently, new polyploids are usually rapidly penalized by natural selection, rendering polyploidy an almost certain evolutionary dead end.[6] But on the rare occasion that a WGD does not spell disaster for an organism, it can serve as a source of evolutionary innovation. Polyploidy, you see, is at the core of much *speciation*, or the creation of new species. Polyploids cannot interbreed with their non-polyploid ancestors due to the numerical incompatibility of their chromosomes and are therefore "forced" to embark on a separate mutational journey, which leads to the development of diverse new species. The WGD underlying the giant African snail's transition from water to land,[7] for example, highlights the ability of polyploids to colonize new habitats, transforming themselves into novel species as they go.

When, exactly, are the normally disastrous WGDs "allowed" to persist and produce new species? To answer this question, let us first examine how polyploidy occurs on a small scale. In multicellular organisms such as humans, polyploidy may be restricted to certain tissues, while the rest of the body is made up of "normal", non-polyploid cells. This phenomenon is known as *endopolyploidy*, and it is the norm in everything from our muscles to our livers.[8]

But why do cells undergo endopolyploidization in the first place? There are two answers to this question. Some endopolyploid cells perform specialized developmental functions,[9] and are therefore integral components of multicellular organisms. Many non-polyploid cells, however, can undergo endopolyploidization in response to stress.[10] In plants, for example, droughts, extreme temperatures, high radiation, soil contamination,[11] and disease[12] can all induce endopolyploidization, while predation[13] and injuries[14] can trigger endopolyploidization in animals.

Converting a small portion of one's cells into polyploids is clearly a stress response mechanism, which means that WGDs are yet another kind of nonrandom mutation. But how does endopolyploidy help organisms survive stressful circumstances? It seems that when WGDs multiply the number of copies of certain genes, their increased expression[15] and/or accelerated neofunctionalization allow for rapid adaptation to dynamic environments.

Let us return to general polyploidy. Stress has also been shown to trigger the WGDs that underly polyploidy,[16] because polyploidy, like endopolyploidy, is advantageous under various stressful conditions.[17] Indeed, although polyploids are outperformed by non-polyploids in stable environments, they often thrive in harsh environments and are more resilient to them.[18] Tetraploid rice, for instance, boasts an increased tolerance to drought and salinity,[19] while tetraploid potatoes enjoy an enhanced resistance to parasites.[20] How do polyploids achieve their superior stress response? The full picture is unclear, but we do know that some polyploid grasses produce heavier seeds,[21] and that higher ploidy levels in wheat are associated with a greater photosynthetic capacity.[22] Taken together, such anecdotal evidence suggests that polyploidy provides organisms with invaluable tricks for surviving in times of environmental upheaval.

To demonstrate how unabashedly nonrandom WGDs can be, let us examine the geographical distribution of polyploids. Polyploids tend to occupy drier habitats than closely related non-polyploids.[23] Moreover, the farther one strays from the equator, the higher the polyploid frequency,[24] indicating that polyploidy is a tried and true method for adapting to cold environments. This, however, does not mean that all kinds of polyploidy are equally useful in extremely cold environments, because the frequency of higher order polyploid plants in the Arctic is correlated with latitude![25]

But what about the temporal distribution of WGDs? In other words, is polyploidy equally likely to arise at different times? Of course not – WGDs are strongly correlated with cataclysmic events, many of which resulted in mass extinctions.[26] This is hardly unexpected, because what are cataclysmic events if not gigantic sources of stress? And given that polyploids are often equipped with survival-boosting features, such as an increased cell size[27] and the ability to reproduce asexually,[28] their chances of making it through mass extinctions are relatively high.

But are WGDs really more likely to occur during mass extinctions? Wouldn't it make more sense if WGDs occurred at a more or less constant rate, but only managed to yield new polyploid species in times of acute stress? This may sound more plausible, but the truth is that unreduced gamete levels rise in response to environmental stress,[29] as demonstrated in various fish, amphibian,[30] and plant species.[31] In fact, fossilized plants from the Triassic–Jurassic Extinction – one of the "Big Five" mass extinctions – are characterized by elevated levels of unreduced pollen,[32] providing further proof for the nonrandomness of WGDs.

Incredibly, the nonrandomness does not stop once a genome has been duplicated. WGDs are often followed by massive gene loss, which explains why even genomes that have undergone multiple duplications are not unusually large. Take the ancestor of all flowering plants, which probably had about six chromosomes.[33] One of its descendants, Mexican cotton, seems to have undergone up to seven WGDs in its evolutionary history, which "should" have resulted in hundreds of chromosomes. In practice, however, Mexican cotton has a modest twenty-six chromosomes,[34] implying that its many WGDs resulted in ephemeral polyploidy.

The post-WGD downsizing process probably begins with the loss of WGD-generated duplicate genes, or *ohnologs*.[35] In yeast, for example, much ohnolog loss has historically occurred immediately after WGDs,[36] eliminating roughly 90% of duplicate gene pairs.[37] Now, if the many kinds of mutations that lead to gene loss were random, we would expect all ohnologs to have an approximately equal chance of being eliminated. But these mutations, too, are nonrandom, and an ohnolog's chances of being eliminated depend on its function. Some kinds of yeast genes, such as those involved in DNA repair, do not last long as ohnologs,[38] while other kinds, such as those involved in regulating transcription, tend to persist in duplicate.[39] Similarly, highly expressed and/or essential plant genes are quickly restored to single-copy status, while less-expressed and/or nonessential plant genes frequently persist as ohnologs.[40]

Amazingly, polyploidal gene loss is nonrandom in an additional, puzzling way. Polyploids, as it turns out, aren't always *autopolyploids*, or polyploids that derive their chromosome sets from a single species. They can also be *allopolyploids*, or polyploids that derive their chromosome sets from two or more distinct species. And if the mixing of genomes from different species sounds preposterous, wait until you hear what happens *after* this unconventional mixing occurs! Like all polyploids, allopolyploids undergo extensive gene loss. But in allopolyploids, genes may be lost in a *progenitor-dependent* manner. This was proven in an illuminating study that analyzed the genomes of a yeast allopolyploid and two plant allopolyploids. After assigning every gene to its progenitor genome of origin, researchers showed that once two genomes are combined in an allopolyploid, genes from one of these genomes are much more likely to be lost.[41]

Allopolyploidy, in other words, can lead to nonrandom mutations that wipe out genes from one progenitor genome in particular. Signs of this *biased fractionation*, as it is known, can still be detected in allopolyploid genomes sixty million years later,[42] though biased fractionation patterns may differ among allopolyploid species. The yeast allopolyploid investigated in the study, for instance, only exhibited biased fractionation early on in its evolution, whereas the two plant allopolyploids managed to maintain their biased fractionation patterns indefinitely.[43] Taken together, these findings indicate that allopolyploids, like all polyploids, are the products of a series of undisputedly nonrandom mutations.

CHAPTER 14

Compensatory Coding

If a gene acquires an advantageous mutation, so traditional neo-Darwinist thought goes, its carrier will be more likely to survive and reproduce. Genes, however, do not exist in a vacuum. What if an advantageous mutation in one gene happens to coincide with a disadvantageous mutation in a different gene? Will their carrier be more or less likely to survive and reproduce? The complexity of this problem grows exponentially in organisms such as the tiny water flea, which boasts an astounding 31,000 or so genes.[1] Nature, it is clear, cannot possibly be as clear-cut as biology textbooks would have us believe. The truth is that advantageous mutations are not always propagated and (mildly) disadvantageous mutations are not always eliminated by natural selection, contradicting the outdated view of evolution as a steady process of advantageous mutation accumulation. A good chunk of evolutionary progress, in fact, revolves around the mitigation of the effects of innumerable slightly disadvantageous mutations via the accumulation of *compensatory mutations*.

Think of it this way: gene families, such as the globin gene family, are created when an ancestral gene is subjected to duplication mutations. Consequently, the different genes in a gene family possess very similar sequences, because although their evolution diverges after duplication, the amino acids in many positions along related proteins tend to remain identical. These stubborn amino acids are often the ones that are vital for proper protein folding, and therefore protein functionality. They rarely vary in different members of a gene family, because their substitution would likely result in dysfunctional proteins. Evolution, then, is not just about how amino acids change over time, but about how *conserved* amino acids struggle to avoid changing over time.

Not all conserved amino acids, however, are equally conserved, because some are more vital than others to proper protein folding and functionality. Indeed, the substitution of *strongly conserved* amino acids is never observed in functional proteins, whereas more *weakly conserved* amino acids are prone to acquiring mildly disadvantageous substitution mutations. Given the high frequency of these disadvantageous mutations, it would seem that natural selection is less concerned with favoring "traditionally" advantageous mutations and more concerned with promoting compensatory mutations, whether these occur in the same protein impacted by a disadvantageous mutation or in a different protein with which it interacts.[2]

Non-lethal disadvantageous mutations, as it turns out, are constantly arising and being canceled out, or at least mitigated, by compensatory mutations.[3] In fact, the greater the severity of disadvantageous mutations, the faster the rate of compensatory evolution![4] But how, exactly, do compensatory mutations reduce the effects of disadvantageous mutations? They may minimize the cell's need for a protein affected by a disadvantageous mutation by, say, recruiting a different protein to perform the damaged protein's role. Alternatively, compensatory mutations may restore the damaged protein's functionality by altering the damaged protein itself,[5] or by adjusting a different protein with which it interacts.[6]

Interestingly, compensatory mutations are rarely precise reversions of disadvantageous mutations.[7] This was demonstrated very elegantly in a study on a virus known as *Bacteriophage T7*, in which researchers engineered the deletion mutation of one key viral gene. Following this calamitous mutation, the mutant virus struggled temporarily, but soon bounced back with the help of compensatory mutations. What did these compensatory mutations look like? Mutations that recreate the deleted gene would seem to be the obvious solution to the virus's woes, but none of those emerged. Instead, most of the compensatory mutations popped up in other key genes,[8] suggesting that compensation is a fragmented, trial-and-error-like process.

The gradual nature of compensatory evolution was elucidated in a study on another virus, *Bacteriophage ϕ6*. To uncover the forces shaping the process of compensatory evolution, researchers induced a highly disadvantageous mutation in Bacteriophage ϕ6's genome and then propagated the mutant virus at a range of population sizes. Thanks to the accumulation of compensatory mutations, the virus eventually recovered regardless of population size, but the *manner* in which it recovered varied greatly. In larger populations, viruses recovered almost perfectly with the help of massive mutational leaps. The smaller populations, on the other hand, were characterized by a partial recovery achieved by more modest mutational steps.[9]

This is not to say that compensatory mutations never fully compensate for disadvantageous mutations. In one enlightening study, over 180 unique mutant strains of baker's yeast were engineered by deleting a different gene in each strain. Following a temporary phase of slow growth, the mutant strains began to recover by acquiring compensatory mutations throughout the genome. Soon afterwards, 68% of them made a near-perfect recovery, but more importantly, some mutant strains began to outperform their non-mutant ancestors,[10] indicating that compensatory mutations can do much more than put out fires.

Sometimes, compensatory mutations are necessary to mitigate the effects of mutations that are less obviously disadvantageous. A good example of a secretly disadvantageous mutation is the acquisition of a bacterial plasmid, which, as we saw in Chapter 4, is by definition a mutation. The acquisition of bacterial plasmids can be immensely beneficial, especially if the acquired plasmids bear gifts such as antibiotic resistance genes. Still, this acquisition has disadvantages, since the bacterium must now invest costly resources in replicating, repairing, and expressing extra genes.[11]

Given these substantial costs, is it any wonder that bacteria tend to lose plasmids that are no longer useful? Plasmids with antibiotic resistance genes, for example, are welcomed in the presence of antibiotics, but in the subsequent absence of antibiotics, they are reduced to being an unnecessary burden on bacteria. Fortunately, this burden can be eased by compensatory mutations on the plasmids themselves or on the bacterial chromosome. By altering the expression of genes found on the plasmids and/or the chromosome, modulating the bacterial plasmid-swapping rate, and promoting the loss of certain plasmid genes, compensatory mutations enable bacteria to hold on to their plasmids for longer, resulting in prolonged antibiotic resistance.[12] This, of course, means that if these bacteria are later re-exposed to antibiotics, they are still resistant. But that's not all – bacteria can also gain antibiotic resistance by acquiring compensatory mutations![13] That's right – compensatory mutations don't just preserve antibiotic resistance; they can also create it *ab initio*.

Compensatory mutations are clearly central to evolution, and the role they play in shaping our genomes may even overshadow that of the adaptive mutations more commonly associated with natural selection. But what about their randomness, or lack thereof? Are compensatory mutations equally likely to, for instance, arise in different positions around the genome? Not at all. Even though a disadvantageous mutation in one gene can be easily offset by a compensatory mutation in another gene, compensatory mutations tend to occur very close to sites of disadvantageous mutations. The average compensatory mutation, as it turns out, occurs within a distance from the site of a disadvantageous mutation that is equivalent to 22% of the affected gene's length. Moreover, a single amino acid in the disadvantageous mutation's vicinity will often undergo multiple compensatory mutations, while amino acids a stone's throw away remain untouched.

Incredibly, for compensatory mutations, distance from a disadvantageous mutation is not just a function of a protein's sequence, but also of a protein's three-dimensional structure. An amino acid located at a great distance from the site of a disadvantageous mutation sequence-wise may find itself in close proximity with this site once the protein folds. In this case it, too, will be very likely to acquire compensatory mutations,[14] dispelling any preconceived notions of mutational randomness.

CHAPTER 15

Arcane Aggregation

As we've seen, once a protein is produced, it folds into a particular three-dimensional structure that is crucial to its functionality. This, however, isn't the only possible outcome for proteins; they can also fold incorrectly, or *misfold*, in which case they are dysfunctional. And to add insult to injury, misfolded proteins are prone to *aggregating*, or accumulating to form unsightly clumps that take up a great deal of valuable intracellular space and poke holes in membranes.[1] Consequently, protein aggregation has been implicated in over forty human diseases, including Alzheimer's and Parkinson's.[2]

Protein aggregation occurs so frequently in cells that specialized aggregation zones have evolved in many species. In bacteria, aggregates can be found in one of the cell's "poles", and following cell division, the nascent cell that ends up with this pole grows much more slowly than its unburdened counterpart.[3] Similarly, protein aggregates in yeast can be found in the Juxtanuclear Quality Control compartment (JUNQ),[4] or in the Insoluble Protein Deposit (IPOD),[5] depending on whether they are destined for degradation or permanent deposition, respectively. Diseased mammalian cells, for their part, have been shown to form structures called *aggresomes*,[6] which can take different shapes.

Interestingly, although aggregation was once thought to be unique to a few disease-associated proteins, it has since emerged that aggregation can befall *all* misfolded proteins.[7] Indeed, even "correctly" folded proteins have been shown to exhibit a small, but significant tendency to aggregate.[8] So why do most proteins never form pathological aggregates? One intriguing study set out to answer this question by analyzing the aggregation kinetics of everything from disease-associated proteins to proteins that only aggregate under artificial conditions. Disease-associated proteins were revealed to be the only ones capable of aggregating rapidly, while the aggregation rate of other proteins was simply too slow to be biologically significant.[9] This is a bewildering discovery, because the ubiquity of protein aggregation and the variety of speeds at which it occurs indicate that aggregation – any aggregation – is somehow beneficial.

But what could proteins possibly gain from forming dysfunctional clumps infuriatingly slowly? The key to this conundrum may lie in another fascinating study, which focused on two unexpected protein aggregate residents: DNA and RNA. That's right, DNA and RNA are present in protein aggregates! And when these seemingly misplaced DNA and RNA molecules were sequenced, an unusually high proportion of the resulting sequences turned out to be *viral*. How can we make sense of this discovery? Viral DNA and RNA may be enigmatically predisposed to binding aggregating proteins, but this seems unlikely. A more plausible explanation is that protein aggregation evolved to sequester viral genomes, thus preventing them from hijacking the cell. Protein aggregates may be, at their core, viral DNA and RNA traps,[10] in which case the diseases stemming from accelerated aggregation are no more than the regrettable byproducts of a vital antiviral mechanism.

Protein aggregation, of course, can still have devastating effects, as in the case of *prion* pathologies such as mad cow disease and kuru. But what, exactly, is a prion, and why does it have such a detrimental impact on our health? A prion typically starts out as a normal, properly folded protein known as *PrPC*. However, if PrPC misfolds, it becomes a prion, which is arguably the most troublesome kind of protein. Prions are dysfunctional and prone to aggregating, but incredibly, that's not the worst of it! The *real* problem with prions is that they induce the misfolding of normal PrPC proteins they encounter, which means that a single prion is capable of setting off a mass *prionization* process. The resulting colossal protein aggregation in mammalian brains yields the untreatable, fatal neurodegenerative conditions known as prion diseases.

Prion diseases are infamously infectious. Kuru, for example, was once exceedingly common among cannibals in Papua New Guinea. Up until the early 1960s, when individuals with kuru died, their relatives would consume their bodies in order to help free their spirits. Needless to say, this practice introduced the prions of the dead into new, healthy bodies, yielding endless cycles of prionization and perpetuating the kuru epidemic.

Infection, however, is not the only way prion diseases arise. In fact, infections only account for 2–5% of prion disease cases, while sporadic prionization causes a terrifying 85–90% of cases. Where do the remaining cases originate? In mutations, of course![11] If the gene encoding PrPC acquires a mutation that causes its product to misfold, cells will churn out prions organically – no infections or spontaneous misfoldings required. And although over sixty different mutations have been shown to yield prions, their occurrence is about as nonrandom as it gets, because just four of those mutations are behind over half of all genetic prion diseases.[12]

Curiously enough, aggregation can also be caused by mutations in proteins that do not aggregate. To understand how this can be true, let us take a minute to elaborate on a wondrous class of proteins known as *chaperones*. At high school parties, chaperones are unfortunate teachers tasked with preventing the occurrence of, well, undesirable interactions. The same is true for chaperone proteins. Most new proteins can easily fold into their signature three-dimensional structures. But when some new proteins fold in three-dimensional space, the amino acids along their length may interact in "wrong" ways, culminating in misfolding. Chaperone proteins help these new proteins attain their "correct" structures by binding and stabilizing them as they fold, thus preventing undesirable interactions between amino acids. They can also help misfolded proteins *refold*, preemptively solving the problem of aggregation. Consequently, deleterious mutations in chaperone genes are sometimes the root cause of protein aggregation.

Let us return to the proteins actually undergoing aggregation. As we've seen, all proteins are at risk of aggregating, but which proteins should the cell take pains to protect from aggregation? And how do the cell's "preferences" in this matter influence the kinds of mutations different proteins are "allowed" to acquire?

For starters, the most abundant proteins have the potential to form the largest aggregates, and should therefore be held back from aggregating at all costs. This explains why throughout evolutionary history the most abundant proteins have acquired an unusually large number of mutations that render proteins less aggregation-prone.[13] It also explains why abundant proteins mutate very slowly, because natural selection has a zero tolerance policy for mutations that may result in the aggregation of abundant proteins.[14] Similarly, proteins that degrade very slowly are risky, because if they aggregate, eliminating their aggregates will prove extremely difficult. Slowly degrading proteins are, accordingly, forced to evolve in ways that minimize their chances of aggregation,[15] thus limiting the kinds of mutations they may acquire.

There are also proteins that bind identical proteins to form *homo-oligomers*, or multiprotein structures composed of many copies of the same protein. These proteins are inherently predisposed to sticking together, and are therefore always dangerously close to aggregating. Consequently, these proteins tend to acquire significantly more anti-aggregation mutations than proteins that do not homo-oligomerize. Last but not least, essential proteins should always be prevented from aggregating, because cells cannot survive if an essential protein is out of order. The importance of maintaining functional essential proteins is, of course, a critical evolutionary constraint, which explains why essential proteins have acquired a shockingly large quantity of anti-aggregation mutations.[16]

How do the mutations that decrease aggregation propensity work? Some of them increase protein solubility.[17] Abundant proteins, you see, have an inherent tendency to clump together, simply because there are so many of them around, and the only way to mitigate this danger is to increase their solubility in the intracellular fluid. Other anti-aggregation mutations increase protein charge, which requires some explaining. Some, but not all amino acids possess a positive or negative charge. In practice, this means that when amino acids come together to form a protein, their charges can be summed to yield the protein's total charge. Now, a protein's charge is inversely correlated with its aggregation propensity,[18] so mutations that swap amino acids for those with a greater charge are particularly advantageous. Finally, some anti-aggregation mutations position specific amino acids in key spots along the protein's sequence. This is effective because certain protein regions are predisposed to aggregating, and placing the "right" amino acids in positions flanking these regions goes a long way towards reducing aggregation.[19]

These "right" amino acids, incidentally, are also key to preventing the aggregation of *elastomeric* proteins, which can undergo significant deformations before bouncing back to their original structures. Why don't elastomeric proteins aggregate when they are deformed? Because they are characterized by a disproportionately high number of certain amino acids which minimize their chances of aggregating.[20] Elastomeric proteins, in other words, preferentially acquire specific substitution mutations to ward off aggregation.

The reverse phenomenon can be observed with hydrophobic amino acids, which cannot interact with water. Stretches of over five consecutive hydrophobic amino acids, as it turns out, facilitate aggregation, and are consequently less common than they would be if each position along a protein's sequence was assigned an amino acid at random.[21]

This tendency to mutate away from hydrophobic clusters is also evident on a grander scale when one compares proteins of different ages. As we've seen, new genes have arisen throughout evolutionary history, which means that some proteins have been around for longer than other proteins and are therefore "older". A protein's age is typically estimated by identifying a similar protein in the most distantly related species possible and then calculating approximately how long ago the two species diverged. Similar proteins may, of course, prove impossible to find in different species. If a protein is the product of a "brand-new" gene, similar proteins will be, by definition, nonexistent. Moreover, if deletion mutations wipe out gene families in many species, an "old" protein in a species spared by these deletions will appear much "newer". Similar proteins can also be hard to identify in cases of extreme divergence, in which related proteins no longer bear a strong resemblance to each other. And if a similar protein does exist, but only in species with yet unsequenced genomes,[22] well, then we're out of luck. But let us assume that similar proteins can be identified and that protein age can be estimated accordingly. What can this tell us about the mutations that befall hydrophobic amino acids?

Proteins cannot dispense with hydrophobic amino acids entirely, because they play an important role in protein folding. Proteins, you see, exist in aqueous fluids, and therefore fold so that their hydrophobic amino acids face inwards, while their *hydrophilic* amino acids, which interact well with water, face outwards. Thus, the hydrophobic amino acids in a protein's core stabilize its structure by refusing to come into contact with the surrounding fluids. So while protein sequences must avoid clusters of hydrophobic amino acids, which promote aggregation, they must also contain hydrophobic amino acids in carefully selected positions to form the protein core.

When researchers compared the sequences of "new" and "old" proteins, they discovered that although "new" proteins contain hydrophobic clusters, they avert aggregation by avoiding the most strongly hydrophobic amino acids. "Old" proteins, on the other hand, incorporate strongly hydrophobic amino acids into their sequences, and manage to do so without increasing their risk of aggregation by carefully positioning them to reduce clustering. Put simply, as "new" proteins evolve into "old" proteins, they mutate to acquire more strongly hydrophobic amino acids while simultaneously moving them around to optimize stability and minimize aggregability.[23]

This is interesting, because we instinctively expect proteins to be characterized by a wide variety of solubilities, charges, and amino acid distributions. But natural selection is clearly strongly opposed to mutations that decrease protein solubility, decrease protein charge, or relocate carefully positioned amino acids, indicating that protein evolution is severely restricted.

CHAPTER 16

Genetic Genesis

The axiom that mutations occur randomly with respect to their consequences is not the only axiom of dubious veracity ever held by biologists. For many years, it was almost universally assumed that genes can only arise from pre-existing genes. The first gene, according to this hypothesis, was an exception to this rule, but ever since *its* appearance, new genes had consistently arisen via the duplication mutations discussed in Chapter 2. The idea that new genes could somehow arise from the non-genic DNA found between genes was, in other words, unfathomable. This, of course, is understandable, because it is extremely difficult to imagine how a gene could be conjured into existence in the absence of a blueprint, or how such an alien gene could hope to fit into complex gene networks that have never encountered anything like it.

Recently, however, evidence has emerged that genes can, and do, arise from non-genic DNA,[1] and may later become essential genes, to boot! These *de novo* genes, as they are now known, are actually surprisingly common in genomes. In fact, approximately one third of the genes in the yeast genome are unrelated to genes in other organisms,[2] implying that they were formed from scratch in yeast DNA. But are random mutations responsible for creating de novo genes from previously "meaningless" sequences? Several de novo gene characteristics suggest otherwise.

De novo genes are typically shorter than "older" genes, but significantly longer than we would expect them to be if their length were determined by random mutations.[3] De novo genes also contain more repetitive sequences than expected,[4] and often overlap with "older" genes,[5] implying the existence of mutational biases. Interestingly, de novo genes are preferentially located on X chromosomes[6] – an unequivocally nonrandom characteristic. Moreover, when it comes to gene expression, de novo gene expression is much more tissue-specific than "older" gene expression, and de novo gene expression *levels* are much lower than those of "older" genes.[7] Finally, the evident nonrandomness of the mutations that yield de novo genes also extends to the circumstances under which they occur. De novo genes appear to arise in response to highly dynamic environments,[8] and are therefore at the heart of much evolutionary innovation.[9]

Needless to say, de novo genes would be entirely insignificant if they could not yield de novo proteins. But once a de novo gene arises, how does it manage to produce a protein? In other words, how does a de novo gene persuade the cell's transcription machinery to transcribe it? This step is surprisingly easy, because a great deal of non-genic DNA is routinely transcribed. In mammals, for example, only about 1.2% of the genome consists of genes, but most of the genome is transcribed anyway.[10] Note that saying that most of the genome is transcribed is not the same as saying that most sequences are always transcribed while a minority of sequences are never transcribed. There are, in fact, no consistently transcription-free genomic regions – virtually all non-genic sequences have been transcribed at some point in our evolutionary history,[11] providing them with a chance to evolve into functional genes. This is especially true in mammalian testes, where transcription appears particularly pervasive.[12] Consequently, the testes boast the highest rates of de novo gene expression in the human body.[13]

In practice, the indiscriminate transcription described earlier yields many short amino acid chains, or *peptides*.[14] Although this may seem counterintuitive, these supposedly nonsensical peptides often readily fold into well-defined three-dimensional structures,[15] suggesting that a respectable protein structure can be easily acquired by most sequences. So pervasive transcription results in the creation of many kinds of *folded* peptides, which are subsequently subjected to natural selection. Put simply, a large number of folded peptides produced by non-genic sequences are tested by selective forces, and may go on to evolve into functional proteins.

But what can these selective forces possibly "look for" in seemingly meaningless sequences of amino acids? Their first priority appears to be preventing aggregation,[16] which, as we saw in Chapter 15, is crucial to cellular health. When potential new proteins arise, natural selection doesn't immediately assess their ability to catalyze chemical reactions, etc. Instead, it just penalizes those that are prone to aggregating, allowing the rest to continue their evolutionary journey.

What does this evolutionary journey have in store for aspiring de novo proteins? Having cleared the aggregation hurdle, the peptides must demonstrate basic binding capabilities. Here, again, we are misled by our intuition, because the ability to bind various molecules is not at all hard to come by. In fact, when researchers screened a multitude of random amino acid sequences, they discovered that many could successfully bind proteins and a variety of other molecules.[17] Although protein binding is traditionally depicted in textbooks as a strong, highly specific interaction between two compatible proteins, multiple weak, less specific bonds often suffice to get the job done. Consequently, many prospective proteins may give binding interactions a shot before natural selection singles out the most useful ones.

How can a de novo protein candidate be useful? This, too, isn't as complicated as it first seems. Take the case of de novo genes that promote cancer progression. While these genes cannot come up with a completely novel way to facilitate uncontrollable cell division, they can easily reactivate many of the existing pathways that cells silence in the course of their lives. Cells may, for instance, cease dividing when it is time for an organ to stop growing because a protein responsible for cell division becomes unstable at just the right time. Now, if a de novo protein happens to bind and stabilize the unstable protein, it will induce excessive cell divisions, resulting in a tumor.

Thus, the de novo gene is "accidentally" assigned a function, allowing it to commence a process of specialization. This must happen often, because de novo genes tend to emerge very frequently. But if their appearance is so common, why isn't the number of genes in a genome constantly growing? Why, in other words, are genomes typically characterized by a relatively constant number of genes? The only conceivable answer to this question is that genes "die" as frequently as they are "born".[18] Loss-of-function mutations are much more likely to occur in de novo genes than in "older" genes,[19] implying that the high rates of de novo gene "birth" are balanced out by high rates of gene "death" that primarily affect the de novo genes themselves.

What does a "dead" gene look like? A "dead" gene, or *pseudogene*, is a nonfunctional segment of DNA that resembles a functional gene, but contains defects that prevent the production of a functional protein. The most common defects in pseudogenes appear to be nonsense mutations and frameshift mutations, which result in the production of truncated or "meaningless" amino acid sequences. These and other defects are sometimes overcome when pseudogenes regain their original function[20] or evolve a brand-new one,[21] though this happens very rarely.

There are three types of pseudogenes. The first type consists of functional genes that have mutated into pseudogenes by losing their ability to produce functional proteins. At first glance, this doesn't make sense, because why would natural selection "allow" this to happen? Wouldn't mutations that remove functional genes from the genome be immediately penalized? In most cases, yes. But sometimes, genes are rendered obsolete, and for these genes, pseudogenization is far from catastrophic. Take the development of color vision, an evolutionary breakthrough that drastically reduced our reliance on other senses, such as olfaction. The development of color vision in our ancestors may have "allowed" multiple olfactory receptor genes to undergo pseudogenization,[22] which is why color-blind species such as dogs are much better at picking up and discerning scents than humans.

The second and most common type of pseudogene consists of genes that were duplicated and subsequently pseudogenized. As we saw in Chapter 2, gene duplication mutations may result in neofunctionalization or subfunctionalization. But there's also a third option – if one copy of a duplicated gene continues to dutifully churn out proteins, there's no harm in the other copy wasting away. Gene duplication, in other words, creates a redundancy that can easily lead to harmless pseudogenization.

The third type of pseudogene is somewhat similar to the second type. Here, genes are duplicated in the course of *retrotransposition*, which means that as the retrotransposons discussed in Chapter 8 copy and paste themselves into new locations in the genome, they may accidentally take nearby genes along for the ride.[23] But since retrotransposons transcribe themselves into RNA before reverse transcribing the RNA back into DNA, the genes they inadvertently copy "lose" untranscribed flanking sequences, which are vital to gene transcription. Thus, these genes lose the ability to produce proteins as soon as they are integrated into a new genomic location. Retrotransposed pseudogenes, incidentally, accumulate mutations faster than other pseudogenes.[24]

Pseudogenes often accumulate in genomes due to the absence of mechanisms for their removal. Still, identifying pseudogenes can prove challenging because the sequence of each suspected pseudogene must be compared with all known protein-coding sequences to evaluate its functionality. At this point, extremely stringent criteria must be employed to minimize the number of slightly defective, but functional genes that are erroneously declared pseudogenes. Only genes with severely deleterious mutations are considered, resulting in the underestimation of the number of "young" pseudogenes, which have yet to acquire major mutations. Similarly, the number of "very old" pseudogenes is often underestimated because these pseudogenes have acquired so many mutations that their sequences rarely bear any resemblance to the sequences of functional genes. Consequently, the human genome's estimated 12,000 to 14,000 pseudogenes are likely just the tip of the pseudogene iceberg.[25]

Are the mutations that create this abundance of pseudogenes random? In some ways, absolutely. Many species are *symbionts*, which means that they live in intricate interdependence with other species, or *parasites*, which means that they live off the fruit of other species' labor. In both cases, a sizable number of genes found in the genomes of *free-living species* are rendered obsolete, because so many of the symbionts' and parasites' physiological needs are met by other species. Consequently, deleterious mutations in these genes are tolerated, and many obsoleted genes end up undergoing pseudogenization. In this case, pseudogenizing mutations can be distributed randomly throughout the genome with very few repercussions. But pseudogenizing mutations also have an emphatically nonrandom side. In barley, for example, pseudogenes tend to be found at the ends of chromosomes and in gene clusters,[26] implying a mutational bias. Moreover, certain kinds of genes, such as those that encode DNA-binding proteins, are very likely to generate many pseudogenes, whereas some kinds of genes don't generate pseudogenes at all.[27]

But what about the mutations acquired *after* pseudogenization? Are they, at least, random? Once a gene becomes a pseudogene, the mutations it accumulates are rendered invisible to natural selection, because they make no difference anyway. Unencumbered by selective constraints, these mutations should be the epitome of randomness. But are they really?

A good way to check is to examine the mutations accumulated by pseudogenes in different species, where their sequences should, theoretically, have free rein to mutate in every conceivable way. Take, for instance, the *L-Gulono-gamma-lactone oxidase*, or *GULO*, enzyme, which is responsible for the last step of vitamin C synthesis. Humans need to consume vitamin C because they cannot synthesize it due to the human GULO gene's pseudogenization. Guinea pigs, too, possess a GULO gene that has undergone pseudogenization, but since most mammals possess a functional GULO gene, this pseudogenization must have occurred completely independently from ours. The human GULO pseudogene and the guinea pig GULO pseudogene "should" therefore have acquired completely different sets of mutations. But, intriguingly, that is not what happened. A comparison of the nucleotide substitutions that have occurred in the two pseudogenes revealed that 36% of the substitution mutations in the human GULO pseudogene were mirrored in the guinea pig GULO pseudogene. This percentage is, of course, too high to have arisen randomly. In fact, the chances of this percentage arising randomly were calculated to be about one in half a trillion,[28] demonstrating that even in pseudogenes, mutations are far from random.

Epilogue

As we saw in Chapter 1, Darwin's grandfather, Erasmus Darwin, developed his own theory of evolution long before the birth of his more famous grandson. But although much has been written about the grandfather's ideas regarding acquired inheritance, his views on the origin of life are rarely discussed. That's a pity, because Erasmus Darwin believed that all life had arisen from shells, and even designed his family's coat of arms accordingly! This coat of arms, which he proudly displayed on his carriage, featured a shield with three exquisite scallop shells and the motto "*E conchis omnia*", or "everything from shells".

Erasmus Darwin clearly realized that all organisms had evolved from a simple ancestor. But what came before the earliest organism? It is all too easy to assume that the *primordial soup*, as the mixture of organic molecules found in the Earth's primitive water bodies is known, was full of free DNA molecules that were eventually enveloped by proto-membranes. However, this hypothesis is lacking, because although DNA can store hereditary information in an easily replicable format, this information is only useful when it can be a blueprint for proteins, which catalyze the chemical reactions required for life. DNA itself cannot catalyze chemical reactions, and it is therefore incapable of catalyzing its own replication. So how did multiple DNA molecules come to inhabit the primordial soup in the first place?

Tackling this question requires yet another rethinking of the definition of mutations. Mutations have been defined in this book as changes to the sequence or composition of a genome. But if we expand this definition to include changes to the *chemical composition* of a genome, a new class of groundbreaking mutations emerges at the dawn of life. To understand what these mutations entailed, we turn to RNA, which is used to convey the hereditary information stored in DNA to the protein-building machinery. This, however, is not all RNA can do. Like DNA, RNA can be used to store hereditary information in an easily replicable format. Some viruses, as we've seen, have RNA genomes, which means that their hereditary information is stored in and replicated on RNA – no DNA required. Similarly, the telomeres, or regions of repetitive sequences at the ends of chromosomes, are elongated during embryonic development via a mechanism that involves replicating an RNA template.[1]

Does this make RNA just another version of DNA? Hardly. Like proteins, RNA can catalyze chemical reactions. RNA enzymes, or *ribozymes*, can even catalyze the replication of their own sequences, in stark contrast to DNA. And incredibly, when researchers artificially synthesized short self-replicating RNA molecules, one of them displayed an error rate of just 1.1% per nucleotide,[2] demonstrating that RNA can replicate itself extremely accurately. In other words, RNA can singlehandedly do everything we rely on DNA and proteins to do, which is why biologists have long suspected that self-replicating RNA preceded the "invention" of DNA.[3] Vestiges of this "RNA world", as it is known, may include *viroids* – small, self-replicating RNA molecules that infect plants. Unlike viruses, viroids lack a protein envelope and do not encode any proteins;[4] these free RNA molecules exist solely to replicate themselves, as all RNA molecules presumably once did.

The presence of amino acids has been shown to increase RNA molecule formation by up to a hundredfold.[5] Moreover, research has revealed that the building blocks of RNA and short amino acid sequences, or peptides, were probably abundant in the primordial soup. Consequently, an alternative "RNA-peptide world" hypothesis has been proposed to allow for greater complexity. According to this hypothesis, peptides used to be assembled directly on primitive RNA molecules. Those peptides may have played a role in stabilizing the RNA molecules, allowing them to evolve longer, more complex sequences,[6] which could encode longer, more complex peptides.

Either way, the ascension of RNA genomes likely led to a great deal of experimentation with novel nucleotides before our ancestral genome settled on a four-nucleotide composition. This, of course, was a unique mutational process, the likes of which no longer exist. In a way, it reflects the process by which alphabets gain and lose letters. The Latin alphabet, for example, gained three new letters by royal decree of the Roman emperor Claudius in the first century BCE. The English alphabet, on the other hand, lost at least five letters over the years, as exemplified by the Old English letter *eth* (Đ/ð), which was replaced by the letter *thorn* (Þ/þ), which, in turn, was replaced by the modern digraph "th". Similarly, early RNA genomes must have gained and lost multiple kinds of nucleotides.[7] Needless to say, each of these gains and losses constituted a highly influential mutation forever lost to (natural) history.

Eventually, the "RNA world" or "RNA-peptide world" gave way to the "DNA-RNA-protein world" we know and love. How did this transition occur? A clue to this mystery may lie in a study that demonstrated how a virus with a DNA genome can acquire a gene from a virus with an RNA genome and reverse transcribe it into DNA.[8] This fascinating mechanism suggests that once DNA was "invented", early prebiotic beings could gradually convert their RNA genes into DNA genes, ushering in a new, DNA-based age.

But how did the RNA that preceded DNA come to be? According to the "PAH world" hypothesis, *polycyclic aromatic hydrocarbons*, or *PAHs*, reigned in a pre-RNA era.[9] PAHs are by far the most abundant large molecules in the universe,[10] and have been detected in everything from meteorites[11] to astronomical nebulae.[12] It is therefore highly likely that they were present in the primordial soup, where they could have easily self-assembled into stacked structures. In stacked PAH structures, the distance between adjacent PAHs is 0.34 nm – a distance identical to the one between adjacent nucleotides of RNA or DNA! This would allow RNA nucleotides to bind stacked PAHs, before binding each other to form RNA chains. Thus, naturally occurring PAH stacks could have served as scaffolds for the formation of the first RNA molecules, giving rise to the "RNA world".

In summary, "genomes" may once have consisted of stacks of PAHs that could only self-assemble – a crude, primitive form of self-replication. If we expand the definition of mutations to include changes to the chemical composition of genomes, then the first major mutation in our evolutionary history was the transition from PAH "genomes" to RNA genomes. This mutation was a very advantageous one, because RNA can store hereditary information, replicate itself, and catalyze reactions. But the resulting "RNA world", or some version thereof, did not last long, evolutionarily speaking. Eventually, the second major mutation in our evolutionary history occurred: the transition from RNA genomes to DNA genomes. This, too, was a very advantageous mutation, because DNA is significantly more stable than RNA, which means that it can form longer chains that store more information while simultaneously minimizing its mutation rate at a much lower energetic cost. The titanic impact of these two ancient mutations is impossible to overstate, but how random was their occurrence? We are unlikely to ever find out.

References

Chapter 1: Lamarckian Lore

1. Kočandrle, R., & Kleisner, K. (2013). Evolution born of moisture: Analogies and parallels between Anaximander's ideas on origin of life and man and later pre-Darwinian and Darwinian evolutionary concepts. *Journal of the History of Biology*, 46, 103–124.
2. Sedley, D. (2016). Empedoclean superorganisms. *Rhizomata*, 4(1), 111–125.
3. Zirkle, C. (1935). The inheritance of acquired characters and the provisional hypothesis of pangenesis. *The American Naturalist*, 69(724), 417–445.
4. Burkhardt, Jr, R. W. (1977). *The Spirit of System: Lamarck and Evolutionary Biology*. Cambridge, MA: Harvard University Press.
5. Dobzhansky, T. (1973). Nothing in biology makes sense except in the light of evolution. *The American Biology Teacher*, March. Online at: https://www.pbs.org/wgbh/evolution/library/10/2/text_pop/l_102_01.html (last accessed 6 May 2026).
6. Zirkle, C. (1935). The inheritance of acquired characters and the provisional hypothesis of pangenesis. *The American Naturalist*, 69(724), 417–445.
7. Barlow, N. (ed.) (1958). *The Autobiography of Charles Darwin 1809–1882*. London: Collins. p. 139. Online at https://darwin-online.org.uk (last accessed 6 May 2026).
8. Darwin, C. R. (1859). *On the Origin of Species*. London: John Murray [1st edition]. p. 131. Online at https://darwin-online.org.uk (last accessed 6 May 2026).

Chapter 2: Mutation Menagerie

1. Kimsey, I. J., Szymanski, E. S., Zahurancik, W. J., Shakya, A., Xue, Y., Chu, C. C., ... & Al-Hashimi, H. M. (2018). Dynamic basis for dG• dT misincorporation via tautomerization and ionization. *Nature*, 554(7691), 195–201.
2. Teather, K. L. (2024). *The Evolution of Sex: Strategies of Males and Females*. Oxford, UK: Oxford University Press. pp. 99–118.
3. Lindahl, T. (1993). Instability and decay of the primary structure of DNA. *Nature*, 362(6422), 709–715.
4. Kim, T. W., Delaney, J. C., Essigmann, J. M., & Kool, E. T. (2005). Probing the active site tightness of DNA polymerase in subangstrom

increments. *Proceedings of the National Academy of Sciences*, 102(44), 15803–15808.
5. Crick, F. H. (1966). Codon–anticodon pairing: The wobble hypothesis. *Journal of Molecular Biology*, 19(2), 548–555.
6. Olson, M. V. (1999). When less is more: Gene loss as an engine of evolutionary change. *The American Journal of Human Genetics*, 64(1), 18–23.
7. Kudla, G., Murray, A. W., Tollervey, D., & Plotkin, J. B. (2009). Coding-sequence determinants of gene expression in *Escherichia coli*. *Science*, 324(5924), 255–258.
8. Rodriguez, A., Diehl, J. D., Wright, G. S., Bonar, C. D., Lundgren, T. J., Moss, M. J., ... & Clark, P. L. (2024). Synonymous codon substitutions modulate transcription and translation of a divergent upstream gene by modulating antisense RNA production. *Proceedings of the National Academy of Sciences*, 121(36), e2405510121.
9. Veit, G., Avramescu, R. G., Chiang, A. N., Houck, S. A., Cai, Z., Peters, K. W., ... & Lukacs, G. L. (2016). From CFTR biology toward combinatorial pharmacotherapy: Expanded classification of cystic fibrosis mutations. *Molecular Biology of the Cell*, 27(3), 424–433.
10. Mickle, J. E., & Cutting, G. R. (1998). Clinical implications of cystic fibrosis transmembrane conductance regulator mutations. *Clinics in Chest Medicine*, 19(3), 443–458.
11. Ijdo, J. W., Baldini, A., Ward, D. C., Reeders, S. T., & Wells, R. A. (1991). Origin of human chromosome 2: An ancestral telomere-telomere fusion. *Proceedings of the National Academy of Sciences*, 88(20), 9051–9055.
12. Poszewiecka, B., Gogolewski, K., Stankiewicz, P., & Gambin, A. (2022). Revised time estimation of the ancestral human chromosome 2 fusion. *BMC Genomics*, 23(6), 1–16.
13. Meyer, M., Kircher, M., Gansauge, M. T., Li, H., Racimo, F., Mallick, S., ... & Pääbo, S. (2012). A high-coverage genome sequence from an archaic Denisovan individual. *Science*, 338(6104), 222–226.
14. Vinogradov, S. N., Hoogewijs, D., Bailly, X., Arredondo-Peter, R., Gough, J., Dewilde, S., ... & Vanfleteren, J. R. (2006). A phylogenomic profile of globins. *BMC Evolutionary Biology*, 6, 1–17.
15. Schneider, T., Tan, Y., Li, H., Fisher, J. S., & Zhang, D. (2022). Photoglobin, a distinct family of non-heme binding globins, defines a potential photosensor in prokaryotic signal transduction systems. *Computational and Structural Biotechnology Journal*, 20, 261–273.
16. Storz, J. F. (2016). Gene duplication and evolutionary innovations in hemoglobin-oxygen transport. *Physiology*, 31(3), 223–232.
17. Vanek, T., & Kohli, A. (2023). Biochemistry, myoglobin. StatPearls [online].
18. Pillai, A. S., Chandler, S. A., Liu, Y., Signore, A. V., Cortez-Romero, C. R., Benesch, J. L., ... & Thornton, J. W. (2020). Origin of complexity in haemoglobin evolution. *Nature*, 581(7809), 480–485.

19. Goodman, M., Moore, G. W., & Matsuda, G. (1975). Darwinian evolution in the genealogy of haemoglobin. *Nature*, 253(5493), 603–608.
20. Pritišanac, E., Urlesberger, B., Schwaberger, B., & Pichler, G. (2021). Fetal hemoglobin and tissue oxygenation measured with near-infrared spectroscopy – a systematic qualitative review. *Frontiers in Pediatrics*, 9, 710465.
21. Morioka, I., Morikawa, S., Yusoff, S., Harahap, I. S. K., Nishimura, N., Yokoyama, N., … & Nishio, H. (2010). Genetic disorders associated with neonatal jaundice. *Eastern Journal of Medicine*, 15(4), 155.
22. Kolawole, E. O., Ayeni, E. T., Abolade, S. A., Ugwu, S. E., Awoyinka, T. B., Ofeh, A. S., & Okolo, B. O. (2023). Malaria endemicity in sub-Saharan Africa: Past and present issues in public health. *Microbes and Infectious Diseases*, 4(1), 242–251.
23. Williams, T. N., Mwangi, T. W., Wambua, S., Alexander, N. D., Kortok, M., Snow, R. W., & Marsh, K. (2005). Sickle cell trait and the risk of *Plasmodium falciparum* malaria and other childhood diseases. *The Journal of Infectious Diseases*, 192(1), 178–186.
24. Dean, L., & Dean, L. (2005). *Blood Groups and Red Cell Antigens* (Vol. 2). Bethesda: NCBI.
25. Tournamille, C., Colin, Y., Cartron, J. P., & Le Van Kim, C. (1995). Disruption of a GATA motif in the Duffy gene promoter abolishes erythroid gene expression in Duffy-negative individuals. *Nature Genetics*, 10(2), 224–228.

Chapter 3: Forgettable Flukes

1. Dolgin, E. (2009, 27 August). Human mutation rate revealed. *Nature*. Online at https://www.nature.com/articles/news.2009.864 (last accessed 16 April 2026).
2. Rennell, D., Bouvier, S. E., Hardy, L. W., & Poteete, A. R. (1991). Systematic mutation of bacteriophage T4 lysozyme. *Journal of Molecular Biology*, 222(1), 67–88.
3. Ségurel, L., Thompson, E. E., Flutre, T., Lovstad, J., Venkat, A., Margulis, S. W., … & Przeworski, M. (2012). The ABO blood group is a trans-species polymorphism in primates. *Proceedings of the National Academy of Sciences*, 109(45), 18493–18498.
4. Lopez, J., & Tait, S. W. G. (2015). Mitochondrial apoptosis: Killing cancer using the enemy within. *British Journal of Cancer*, 112(6), 957–962.
5. Milholland, B., Dong, X., Zhang, L., Hao, X., Suh, Y., & Vijg, J. (2017). Differences between germline and somatic mutation rates in humans and mice. *Nature Communications*, 8(1), 15183.
6. Burnet, F. M. (1974). Intrinsic mutagenesis: A genetic basis of ageing. *Pathology*, 6(1), 1–11.
7. Robinson, P. S., Coorens, T. H., Palles, C., Mitchell, E., Abascal, F., Olafsson, S., … & Stratton, M. R. (2021). Increased somatic mutation burdens in normal human cells due to defective DNA polymerases. *Nature Genetics*, 53(10), 1434–1442.

8. Dolgin, E. (2017). The most popular genes in the human genome. *Nature*, 551(7681), 427–432.
9. Simpson, K. J., Mian, S., Forrestel, E. J., Hackel, J., Morton, J. A., Leitch, A. R., & Leitch, I. J. (2024). Bigger genomes provide environment-dependent growth benefits in grasses. *New Phytologist*, 244(5), 2049–2061.
10. Johnston, M. (2000). The yeast genome: On the road to the Golden Age. *Current Opinion in Genetics & Development*, 10(6), 617–623.
11. Goebl, M. G., & Petes, T. D. (1986). Most of the yeast genomic sequences are not essential for cell growth and division. *Cell*, 46(7), 983–992.
12. Suresh, H., Crow, M., Jorstad, N., Hodge, R., Lein, E., Dobin, A., ... & Gillis, J. (2023). Comparative single-cell transcriptomic analysis of primate brains highlights human-specific regulatory evolution. *Nature Ecology & Evolution*, 7(11), 1930–1943.

Chapter 4: Predictable Patterns

1. Wichman, H. A., Badgett, M. R., Scott, L. A., Boulianne, C. M., & Bull, J. J. (1999). Different trajectories of parallel evolution during viral adaptation. *Science*, 285(5426), 422–424.
2. Crandall, K. A., Kelsey, C. R., Imamichi, H., Lane, H. C., & Salzman, N. P. (1999). Parallel evolution of drug resistance in HIV: Failure of nonsynonymous/synonymous substitution rate ratio to detect selection. *Molecular Biology and Evolution*, 16(3), 372–382.
3. Boucher, C. A., O'Sullivan, E., Mulder, J. W., Ramautarsing, C., Kellam, P., Darby, G., ... & Larder, B. A. (1992). Ordered appearance of zidovudine resistance mutations during treatment of 18 human immunodeficiency virus-positive subjects. *Journal of Infectious Diseases*, 165(1), 105–110.
4. Raeside, C., Gaffé, J., Deatherage, D. E., Tenaillon, O., Briska, A. M., Ptashkin, R. N., ... & Schneider, D. (2014). Large chromosomal rearrangements during a long-term evolution experiment with *Escherichia coli*. *mBio*, 5(5), e01377–14.
5. Pupo, G. M., Lan, R., & Reeves, P. R. (2000). Multiple independent origins of *Shigella* clones of *Escherichia coli* and convergent evolution of many of their characteristics. *Proceedings of the National Academy of Sciences*, 97(19), 10567–10572.
6. Bianconi, E., Piovesan, A., Facchin, F., Beraudi, A., Casadei, R., Frabetti, F., ... & Canaider, S. (2013). An estimation of the number of cells in the human body. *Annals of Human Biology*, 40(6), 463–471.
7. Blount, Z. D., Borland, C. Z., & Lenski, R. E. (2008). Historical contingency and the evolution of a key innovation in an experimental population of *Escherichia coli*. *Proceedings of the National Academy of Sciences*, 105(23), 7899–7906.
8. Shtarkman, Y. M., Koçer, Z. A., Edgar, R., Veerapaneni, R. S., D'Elia, T., Morris, P. F., & Rogers, S. O. (2013). Subglacial Lake Vostok (Antarctica) accretion ice contains a diverse set of sequences from

aquatic, marine and sediment-inhabiting bacteria and eukarya. *PLoS One*, 8(7), e67221.

9. Melamed, D., Nov, Y., Malik, A., Yakass, M. B., Bolotin, E., Shemer, R., ... & Livnat, A. (2022). De novo mutation rates at the single-mutation resolution in a human HBB gene region associated with adaptation and genetic disease. *Genome Research*, 32(3), 488–498.
10. Chan, J. A., Fowkes, F. J., & Beeson, J. G. (2014). Surface antigens of *Plasmodium falciparum*-infected erythrocytes as immune targets and malaria vaccine candidates. *Cellular and Molecular Life Sciences*, 71, 3633–3657.
11. Claessens, A., Hamilton, W. L., Kekre, M., Otto, T. D., Faizullabhoy, A., Rayner, J. C., & Kwiatkowski, D. (2014). Generation of antigenic diversity in *Plasmodium falciparum* by structured rearrangement of *Var* genes during mitosis. *PLoS Genetics*, 10(12), e1004812.
12. Chahwan, R., Edelmann, W., Scharff, M. D., & Roa, S. (2012, August). AIDing antibody diversity by error-prone mismatch repair. *Seminars in Immunology*, 24(4), 293–300.
13. Sender, R., Weiss, Y., Navon, Y., Milo, I., Azulay, N., Keren, L., ... & Milo, R. (2023). The total mass, number, and distribution of immune cells in the human body. *Proceedings of the National Academy of Sciences*, 120(44), e2308511120.
14. Hille, F., Richter, H., Wong, S. P., Bratovič, M., Ressel, S., & Charpentier, E. (2018). The biology of CRISPR-Cas: Backward and forward. *Cell*, 172(6), 1239–1259.
15. Fitzpatrick, D. A. (2012). Horizontal gene transfer in fungi. *FEMS Microbiology Letters*, 329(1), 1–8.
16. Hotopp, J. C. D. (2011). Horizontal gene transfer between bacteria and animals. *Trends in Genetics*, 27(4), 157–163.
17. Inoue, Y. (1997). Spontaneous loss of antibiotic-resistant plasmids transferred to *Escherichia coli* in experimental chronic bladder infection. *International Journal of Urology*, 4(3), 285–288.
18. Schartl, M., & Lamatsch, D. K. (2023). How to manage without a Y chromosome. *Proceedings of the National Academy of Sciences*, 120(2), e2218839120.
19. Boto, L. (2010). Horizontal gene transfer in evolution: Facts and challenges. *Proceedings of the Royal Society B: Biological Sciences*, 277(1683), 819–827.
20. Woese, C. R. (2002). On the evolution of cells. *Proceedings of the National Academy of Sciences*, 99(13), 8742–8747.
21. Petes, T. D., Byers, B., & Fangman, W. L. (1973). Size and structure of yeast chromosomal DNA. *Proceedings of the National Academy of Sciences*, 70(11), 3072–3076.
22. Sinclair, D. A., & Guarente, L. (1997). Extrachromosomal rDNA circles – a cause of aging in yeast. *Cell*, 91(7), 1033–1042
23. Schenkel, L., Wang, X., Le, N., Burger, M., & Kroschewski, R. (2023). A dedicated cytoplasmic container collects extrachromosomal DNA away from the mammalian nucleus. *Molecular Biology of the Cell*, 34(11), ar105.

24. Brinton Jr, C. C. (1965). The structure, function, synthesis and genetic control of bacterial pili and a molecular model for DNA and RNA transport in gram negative bacteria. *Transactions of the New York Academy of Sciences*, 27(8, Series II), 1003–1054.
25. Jain, R., Rivera, M. C., & Lake, J. A. (1999). Horizontal gene transfer among genomes: The complexity hypothesis. *Proceedings of the National Academy of Sciences*, 96(7), 3801–3806.
26. Press, M. O., Queitsch, C., & Borenstein, E. (2016). Evolutionary assembly patterns of prokaryotic genomes. *Genome Research*, 26(6), 826–833.
27. Groussin, M., Poyet, M., Sistiaga, A., Kearney, S. M., Moniz, K., Noel, M., ... & Alm, E. J. (2021). Elevated rates of horizontal gene transfer in the industrialized human microbiome. *Cell*, 184(8), 2053–2067.
28. Haskett, T. L., Terpolilli, J. J., Bekuma, A., O'Hara, G. W., Sullivan, J. T., Wang, P., ... & Ramsay, J. P. (2016). Assembly and transfer of tripartite integrative and conjugative genetic elements. *Proceedings of the National Academy of Sciences*, 113(43), 12268–12273.
29. Martin, W., & Mentel, M. (2010). The origin of mitochondria. *Nature Education*, 3(9), 58.
30. Burger, G., Gray, M. W., & Lang, B. F. (2003). Mitochondrial genomes: Anything goes. *Trends in Genetics*, 19(12), 709–716.
31. Tsuji, J., Frith, M. C., Tomii, K., & Horton, P. (2012). Mammalian NUMT insertion is non-random. *Nucleic Acids Research*, 40(18), 9073–9088.
32. Zhou, W., Karan, K. R., Gu, W., Klein, H. U., Sturm, G., De Jager, P. L., ... & Mills, R. E. (2024). Somatic nuclear mitochondrial DNA insertions are prevalent in the human brain and accumulate over time in fibroblasts. *PLoS Biology*, 22(8), e3002723.
33. Tang, S., Conte, V., Zhang, D. J., Žedaveinytė, R., Lampe, G. D., Wiegand, T., ... & Sternberg, S. H. (2024). De novo gene synthesis by an antiviral reverse transcriptase. *Science*, 386(6717), eadq0876.

Chapter 5: Taxonomic Tangles

1. Kinene, T., Wainaina, J., Maina, S., & Boykin, L. M. (2016). Rooting trees, methods for. In R. M. Kliman (ed.), *Encyclopedia of Evolutionary Biology* (pp. 489–493). Oxford, UK: Academic Press (Elsevier).
2. Elliott, M. J., Knerr, N. J., & Schmidt-Lebuhn, A. N. (2018). Choice between phylogram and chronogram can have a dramatic impact on the location of phylogenetic diversity hotspots. *Journal of Biogeography*, 45(9), 2190–2201.
3. Wiley, E. O. (1979). Cladograms and phylogenetic trees. *Systematic Zoology*, 28(1), 88–92.
4. Patterson, N., Richter, D. J., Gnerre, S., Lander, E. S., & Reich, D. (2006). Genetic evidence for complex speciation of humans and chimpanzees. *Nature*, 441(7097), 1103–1108.
5. Rokas, A., & Carroll, S. B. (2006). Bushes in the tree of life. *PLoS Biology*, 4(11), e352.

6. Abouheif, E., Zardoya, R., & Meyer, A. (1998). Limitations of metazoan 18S rRNA sequence data: Implications for reconstructing a phylogeny of the animal kingdom and inferring the reality of the Cambrian explosion. *Journal of Molecular Evolution*, 47, 394–405.
7. Flessa, K. W., Erben, H. K., Hallam, A., Hsü, K. J., Hüssner, H. M., Jablonski, D., ... & Vermeij, G. J. (1986). Causes and consequences of extinction: Group report. In D. M. Raup & D. Jablonski (eds), *Patterns and Processes in the History of Life: Report of the Dahlem Workshop on Patterns and Processes in the History of Life Berlin 1985, June 16–21* (pp. 234–257). Berlin, Heidelberg: Springer Berlin Heidelberg.
8. Wignall, P. B., & Benton, M. J. (1999). Lazarus taxa and fossil abundance at times of biotic crisis. *Journal of the Geological Society*, 156(3), 453–456.
9. Tenaillon, O., Rodríguez-Verdugo, A., Gaut, R. L., McDonald, P., Bennett, A. F., Long, A. D., & Gaut, B. S. (2012). The molecular diversity of adaptive convergence. *Science*, 335(6067), 457–461.
10. Hug, S. M., & Gaut, B. S. (2016). Lazarus effects: The frequency and genetic causes of *Escherichia coli* population recovery under lethal heat stress. *bioRxiv*, 058479.

Chapter 6: Biased Bursts

1. Drake, J. W., Charlesworth, B., Charlesworth, D., & Crow, J. F. (1998). Rates of spontaneous mutation. *Genetics*, 148(4), 1667–1686.
2. Soojin, V. Y. (2013). Morris Goodman's hominoid rate slowdown: The importance of being neutral. *Molecular Phylogenetics and Evolution*, 66(2), 569–574.
3. Elango, N., Thomas, J. W., NISC Comparative Sequencing Program, & Yi, S. V. (2006). Variable molecular clocks in hominoids. *Proceedings of the National Academy of Sciences*, 103(5), 1370–1375.
4. Scally, A., & Durbin, R. (2012). Revising the human mutation rate: Implications for understanding human evolution. *Nature Reviews Genetics*, 13(10), 745–753.
5. Holmes, E. C. (2011). What does virus evolution tell us about virus origins? *Journal of Virology*, 85(11), 5247–5251.
6. Martin, A. P., & Palumbi, S. R. (1993). Body size, metabolic rate, generation time, and the molecular clock. *Proceedings of the National Academy of Sciences*, 90(9), 4087–4091.
7. Davies, T. J., Savolainen, V., Chase, M. W., Moat, J., & Barraclough, T. G. (2004). Environmental energy and evolutionary rates in flowering plants. *Proceedings of the Royal Society B: Biological Sciences*, 271(1553), 2195–2200.
8. Waldvogel, A. M., & Pfenninger, M. (2021). Temperature dependence of spontaneous mutation rates. *Genome Research*, 31(9), 1582–1589.
9. Denamur, E., & Matic, I. (2006). Evolution of mutation rates in bacteria. *Molecular Microbiology*, 60(4), 820–827.
10. Swings, T., Van den Bergh, B., Wuyts, S., Oeyen, E., Voordeckers, K., Verstrepen, K. J., ... & Michiels, J. (2017). Adaptive tuning of mutation

rates allows fast response to lethal stress in *Escherichia coli. eLife*, 6, e22939.
11. Wei, W., Ho, W. C., Behringer, M. G., Miller, S. F., Bcharah, G., & Lynch, M. (2022). Rapid evolution of mutation rate and spectrum in response to environmental and population-genetic challenges. *Nature Communications*, 13(1), 4752.
12. Pal, C., Maciá, M. D., Oliver, A., Schachar, I., & Buckling, A. (2007). Coevolution with viruses drives the evolution of bacterial mutation rates. *Nature*, 450(7172), 1079–1081.
13. Fitzgerald, D. M., & Rosenberg, S. M. (2019). What is mutation? A chapter in the series: How microbes "jeopardize" the modern synthesis. *PLoS Genetics*, 15(4), e1007995.
14. McKenzie, G. J., Lee, P. L., Lombardo, M. J., Hastings, P. J., & Rosenberg, S. M. (2001). SOS mutator DNA polymerase IV functions in adaptive mutation and not adaptive amplification. *Molecular Cell*, 7(3), 571–579.
15. Uneke, C. J. (2007). Impact of placental *Plasmodium falciparum* malaria on pregnancy and perinatal outcome in sub-Saharan Africa: I: Introduction to placental malaria. *The Yale Journal of Biology and Medicine*, 80(2), 39.
16. Greenspoon, P. B., & M'Gonigle, L. K. (2013). The evolution of mutation rate in an antagonistic coevolutionary model with maternal transmission of parasites. *Proceedings of the Royal Society B: Biological Sciences*, 280(1761), 20130647.
17. Allio, R., Donega, S., Galtier, N., & Nabholz, B. (2017). Large variation in the ratio of mitochondrial to nuclear mutation rate across animals: Implications for genetic diversity and the use of mitochondrial DNA as a molecular marker. *Molecular Biology and Evolution*, 34(11), 2762–2772.
18. Milholland, B., Dong, X., Zhang, L., Hao, X., Suh, Y., & Vijg, J. (2017). Differences between germline and somatic mutation rates in humans and mice. *Nature Communications*, 8(1), 15183.
19. Holmquist, G. P., & Filipski, J. (1994). Organization of mutations along the genome: A prime determinant of genome evolution. *Trends in Ecology & Evolution*, 9(2), 65–69.
20. Hodgkinson, A., & Eyre-Walker, A. (2011). Variation in the mutation rate across mammalian genomes. *Nature Reviews Genetics*, 12(11), 756–766.
21. Wang, G., Fu, L., Miao, W., & Zhang, J. (2025). Linear covariation between germline and somatic mutation rates across ciliates and mammals. *Current Biology*, 35(18):4553–4560.
22. Park, C., Qian, W., & Zhang, J. (2012). Genomic evidence for elevated mutation rates in highly expressed genes. *EMBO Reports*, 13(12), 1123–1129.
23. Monroe, J. G., Srikant, T., Carbonell-Bejerano, P., Becker, C., Lensink, M., Exposito-Alonso, M., ... & Weigel, D. (2022). Mutation bias reflects natural selection in *Arabidopsis thaliana. Nature*, 602(7895), 101–105.
24. Capra, J. A., Erwin, G. D., McKinsey, G., Rubenstein, J. L., & Pollard, K. S. (2013). Many human accelerated regions are developmental enhancers. *Philosophical Transactions of the Royal Society B: Biological Sciences*, 368(1632), 20130025.

25. Wyckoff, G. J., Malcom, C. M., Vallender, E. J., & Lahn, B. T. (2005). A highly unexpected strong correlation between fixation probability of nonsynonymous mutations and mutation rate. *Trends in Genetics*, 21(7), 381–385.
26. Bae, T., Tomasini, L., Mariani, J., Zhou, B., Roychowdhury, T., Franjic, D., ... & Vaccarino, F. M. (2018). Different mutational rates and mechanisms in human cells at pregastrulation and neurogenesis. *Science*, 359(6375), 550–555.
27. Bjedov, I., Tenaillon, O., Gerard, B., Souza, V., Denamur, E., Radman, M., ... & Matic, I. (2003). Stress-induced mutagenesis in bacteria. *Science*, 300(5624), 1404–1409.
28. Harris, K. (2015). Evidence for recent, population-specific evolution of the human mutation rate. *Proceedings of the National Academy of Sciences*, 112(11), 3439–3444.

Chapter 7: Strategic Spots

1. Martincorena, I., Roshan, A., Gerstung, M., Ellis, P., Van Loo, P., McLaren, S., ... & Campbell, P. J. (2015). High burden and pervasive positive selection of somatic mutations in normal human skin. *Science*, 348(6237), 880–886.
2. Campbell, C. D., Chong, J. X., Malig, M., Ko, A., Dumont, B. L., Han, L., ... & Eichler, E. E. (2012). Estimating the human mutation rate using autozygosity in a founder population. *Nature Genetics*, 44(11), 1277–1281.
3. Szczepanik, D., Mackiewicz, P., Kowalczuk, M., Gierlik, A., Nowicka, A., Dudek, M. R., & Cebrat, S. (2001). Evolution rates of genes on leading and lagging DNA strands. *Journal of Molecular Evolution*, 52, 426–433.
4. Fix, D. F., & Glickman, B. W. (1987). Asymmetric cytosine deamination revealed by spontaneous mutational specificity in an Ung$^-$ strain of *Escherichia coli*. *Molecular and General Genetics: MGG*, 209, 78–82.
5. Paul, S., Million-Weaver, S., Chattopadhyay, S., Sokurenko, E., & Merrikh, H. (2013). Accelerated gene evolution through replication – transcription conflicts. *Nature*, 495(7442), 512–515.
6. Lang, G. I., & Murray, A. W. (2011). Mutation rates across budding yeast chromosome VI are correlated with replication timing. *Genome Biology and Evolution*, 3, 799–811.
7. Yehuda, Y., Blumenfeld, B., Mayorek, N., Makedonski, K., Vardi, O., Cohen-Daniel, L., ... & Simon, I. (2018). Germline DNA replication timing shapes mammalian genome composition. *Nucleic Acids Research*, 46(16), 8299–8310.
8. Stamatoyannopoulos, J. A., Adzhubei, I., Thurman, R. E., Kryukov, G. V., Mirkin, S. M., & Sunyaev, S. R. (2009). Human mutation rate associated with DNA replication timing. *Nature Genetics*, 41(4), 393–395.
9. Hu, X. P., Brahmantio, B., Bartoszek, K., & Lercher, M. J. (2025). Most bacterial gene families are biased toward specific chromosomal positions. *Science*, 388(6743), 186–191.

10. Foster, P. L., Hanson, A. J., Lee, H., Popodi, E. M., & Tang, H. (2013). On the mutational topology of the bacterial genome. *G3: Genes, Genomes, Genetics*, 3(3), 399–407.
11. Farr, A. D., Vasileiou, C., Lind, P. A., & Rainey, P. B. (2025). An extreme mutational hotspot in nlpD depends on transcriptional induction of rpoS. *PLoS Genetics*, 21(1), e1011572.
12. Chuang, J. H., & Li, H. (2004). Functional bias and spatial organization of genes in mutational hot and cold regions in the human genome. *PLoS Biology*, 2(2), e29.
13. Todd, P. A., & Glickman, B. W. (1982). Mutational specificity of UV light in *Escherichia coli*: Indications for a role of DNA secondary structure. *Proceedings of the National Academy of Sciences*, 79(13), 4123–4127.
14. Ha, S. C., Lowenhaupt, K., Rich, A., Kim, Y.-G., & Kim, K. K. (2005). Crystal structure of a junction between B-DNA and Z-DNA reveals two extruded bases. *Nature*, 437(7062), 1183–1186.
15. Zhao, J., Bacolla, A., Wang, G., & Vasquez, K. M. (2010). Non-B DNA structure-induced genetic instability and evolution. *Cellular and Molecular Life Sciences*, 67, 43–62.
16. Warburton, P. E., Waye, J. S., & Willard, H. F. (1993). Nonrandom localization of recombination events in human alpha satellite repeat unit variants: Implications for higher-order structural characteristics within centromeric heterochromatin. *Molecular and Cellular Biology*, 13(10), 6520–6529.
17. Ross, D. W. (1999). The human genome: Information content and structure. *Hospital Practice*, 34(6), 49–65.
18. Makova, K. D., & Hardison, R. C. (2015). The effects of chromatin organization on variation in mutation rates in the genome. *Nature Reviews Genetics*, 16(4), 213–223.
19. Chen, X., Chen, Z., Chen, H., Su, Z., Yang, J., Lin, F., ... & He, X. (2012). Nucleosomes suppress spontaneous mutations base-specifically in eukaryotes. *Science*, 335(6073), 1235–1238.
20. Sala, A., Labrador, M., Buitrago, D., De Jorge, P., Battistini, F., Heath, I. B., & Orozco, M. (2024). An integrated machine-learning model to predict nucleosome architecture. *Nucleic Acids Research*, 52(17), 10132–10143.
21. de Koning, A. J., Gu, W., Castoe, T. A., Batzer, M. A., & Pollock, D. D. (2011). Repetitive elements may comprise over two-thirds of the human genome. *PLoS Genetics*, 7(12), e1002384.
22. Haerty, W., & Golding, G. B. (2010). Genome-wide evidence for selection acting on single amino acid repeats. *Genome Research*, 20(6), 755–760.
23. Albertini, A. M., Hofer, M., Calos, M. P., & Miller, J. H. (1982). On the formation of spontaneous deletions: The importance of short sequence homologies in the generation of large deletions. *Cell*, 29(2), 319–328. Streisinger, K. (1966). Frameshift mutations and the genetic code. *Cold Spring Harbor Symposia on Quantitative Biology*, 31, 77–84.
24. Gibson, J. L., Lombardo, M. J., Thornton, P. C., Hu, K. H., Galhardo, R. S., Beadle, B., ... & Rosenberg, S. M. (2010). The σE stress response

is required for stress-induced mutation and amplification in *Escherichia coli*. *Molecular Microbiology*, 77(2), 415–430.

25. Ponder, R. G., Fonville, N. C., & Rosenberg, S. M. (2005). A switch from high-fidelity to error-prone DNA double-strand break repair underlies stress-induced mutation. *Molecular Cell*, 19(6), 791–804.
26. Bull, H. J., McKenzie, G. J., Hastings, P. J., & Rosenberg, S. M. (2000). Evidence that stationary-phase hypermutation in the *Escherichia coli* chromosome is promoted by recombination. *Genetics*, 154(4), 1427–1437.
27. Weigmann, K. (2014). Lifestyle in the sperm: There is growing evidence that epigenetic marks can be inherited. But what is the nature of the information they store and over how many generations do they prevail? *EMBO Reports*, 15(12), 1233–1237.
28. Wu, H. A. O., & Sun, Y. E. (2006). Epigenetic regulation of stem cell differentiation. *Pediatric Research*, 59(4), 21–25.
29. Owen, J. A., Osmanović, D., & Mirny, L. (2023). Design principles of 3D epigenetic memory systems. *Science*, 382(6672), eadg3053.
30. Reik, W., & Walter, J. (2001). Genomic imprinting: Parental influence on the genome. *Nature Reviews Genetics*, 2(1), 21–32.
31. Bernstein, C., & Bernstein, H. (2015). Epigenetic reduction of DNA repair in progression to gastrointestinal cancer. *World Journal of Gastrointestinal Oncology*, 7(5), 30.
32. Hodgkinson, A., & Eyre-Walker, A. (2011). Variation in the mutation rate across mammalian genomes. *Nature Reviews Genetics*, 12(11), 756–766.
33. Coulondre, C., Miller, J. H., Farabaugh, P. J., & Gilbert, W. (1978). Molecular basis of base substitution hotspots in *Escherichia coli*. *Nature*, 274(5673), 775–780.
34. Lynch, M. (2010). Rate, molecular spectrum, and consequences of human mutation. *Proceedings of the National Academy of Sciences*, 107(3), 961–968.
35. Hershberg, R., & Petrov, D. A. (2010). Evidence that mutation is universally biased towards AT in bacteria. *PLoS Genetics*, 6(9), e1001115.
36. Fryxell, K. J., & Moon, W. J. (2005). CpG mutation rates in the human genome are highly dependent on local GC content. *Molecular Biology and Evolution*, 22(3), 650–658.
37. Qu, W., Hashimoto, S. I., Shimada, A., Nakatani, Y., Ichikawa, K., Saito, T. L., & Morishita, S. (2012). Genome-wide genetic variations are highly correlated with proximal DNA methylation patterns. *Genome Research*, 22(8), 1419–1425.

Chapter 8: Viral Vagabonds

1. Schlub, T. E., Grimm, A. J., Smyth, R. P., Cromer, D., Chopra, A., Mallal, S., ... & Davenport, M. P. (2014). Fifteen to twenty percent of HIV substitution mutations are associated with recombination. *Journal of Virology*, 88(7), 3837–3849

2. Szyszka, T. N., Andreas, M. P., Lie, F., Miller, L. M., Adamson, L. S., Fatehi, F., ... & Lau, Y. H. (2024). Point mutation in a virus-like capsid drives symmetry reduction to form tetrahedral cages. *bioRxiv*, 2024–02.
3. Rice, A. M., Castillo Morales, A., Ho, A. T., Mordstein, C., Mühlhausen, S., Watson, S., ... & Hurst, L. D. (2021). Evidence for strong mutation bias toward, and selection against, U content in SARS-CoV-2: Implications for vaccine design. *Molecular Biology and Evolution*, 38(1), 67–83.
4. Niederer, H. A., Laydon, D. J., Melamed, A., Elemans, M., Asquith, B., Matsuoka, M., & Bangham, C. R. (2014). HTLV-1 proviral integration sites differ between asymptomatic carriers and patients with HAM/TSP. *Virology Journal*, 11, 1–12.
5. Albanese, A., Arosio, D., Terreni, M., & Cereseto, A. (2008). HIV-1 pre-integration complexes selectively target decondensed chromatin in the nuclear periphery. *PLoS One*, 3(6), e2413.
6. Leclercq, I., Mortreux, F., Gabet, A. S., Jonsson, C. B., & Wattel, E. (2000). Basis of HTLV type 1 target site selection. *AIDS Research and Human Retroviruses*, 16(16), 1653–1659.
7. Samulski, R. J., Zhu, X., Xiao, X., Brook, J. D., Housman, D. E., Epstein, N. A., & Hunter, L. A. (1991). Targeted integration of adeno-associated virus (AAV) into human chromosome 19. *The EMBO Journal*, 10(12), 3941–3950.
8. LaFave, M. C., Varshney, G. K., Gildea, D. E., Wolfsberg, T. G., Baxevanis, A. D., & Burgess, S. M. (2014). MLV integration site selection is driven by strong enhancers and active promoters. *Nucleic Acids Research*, 42(7), 4257–4269.
9. Chou, K. S., Okayama, A., Su, I. J., Lee, T. H., & Essex, M. (1996). Preferred nucleotide sequence at the integration target site of human T-cell leukemia virus type I from patients with adult T-cell leukemia. *International Journal of Cancer*, 65(1), 20–24
10. McLysaght, A., & Hurst, L. D. (2016). Open questions in the study of de novo genes: What, how and why. *Nature Reviews Genetics*, 17(9), 567–578.
11. Lander, E. S., Linton, L. M., Birren, B., Nusbaum, C., Zody, M. C., Baldwin, J., ... & International Human Genome Sequencing Consortium (2001). Initial sequencing and analysis of the human genome. *Nature*, 409(6822), 860–921.
12. SanMiguel, P., Tikhonov, A., Jin, Y. K., Motchoulskaia, N., Zakharov, D., Melake-Berhan, A., ... & Bennetzen, J. L. (1996). Nested retrotransposons in the intergenic regions of the maize genome. *Science*, 274(5288), 765–768.
13. Gao, C., Xiao, M., Ren, X., Hayward, A., Yin, J., Wu, L., ... & Li, J. (2012). Characterization and functional annotation of nested transposable elements in eukaryotic genomes. *Genomics*, 100(4), 222–230.
14. Liu, C., Zhang, Y., Liu, C. C., & Schatz, D. G. (2022). Structural insights into the evolution of the RAG recombinase. *Nature Reviews Immunology*, 22(6), 353–370.

15. Autexier, C., & Lue, N. F. (2006). The structure and function of telomerase reverse transcriptase. *Annual Review of Biochemistry*, 75(1), 493–517.
16. Brosius, J., & Tiedge, H. (1995). Reverse transcriptase: Mediator of genomic plasticity. *Virus Genes*, 11(2–3): 163–179.
17. Specchia, V., Piacentini, L., Tritto, P., Fanti, L., D'Alessandro, R., Palumbo, G., ... & Bozzetti, M. P. (2010). Hsp90 prevents phenotypic variation by suppressing the mutagenic activity of transposons. *Nature*, 463(7281), 662–665.
18. Prizak, R., Gadzekpo, A., & Hilbert, L. (2025). Chromatin unfolding via loops can drive clustered transposon insertion. *Biophysical Journal*, 124(22), 3908–3922.
19. Wichman, H. A., Van Den Bussche, R. A., Hamilton, M. J., & Baker, R. J. (1993). Transposable elements and the evolution of genome organization in mammals. *Transposable Elements and Evolution*, 149–157.
20. Betrán, E., Thornton, K., & Long, M. (2002). Retroposed new genes out of the X in Drosophila. *Genome Research*, 12(12), 1854–1859.
21. Kazazian Jr, H. H. (2004). Mobile elements: Drivers of genome evolution. *Science*, 303(5664), 1626–1632.
22. Wei, B., Liu, H., Liu, X., Xiao, Q., Wang, Y., Zhang, J., ... & Huang, Y. (2016). Genome-wide characterization of non-reference transposons in crops suggests non-random insertion. *BMC Genomics*, 17, 1–13.
23. Capel, J., Montero, L. M., Martinez-Zapater, J. M., & Salinas, J. (1993). Non-random distribution of transposable elements in the nuclear genome of plants. *Nucleic Acids Research*, 21(10), 2369–2373.
24. Bernardi, G. (1989). The isochore organization of the human genome. *Annual Review of Genetics*, 23(1), 637–659.

Chapter 9: Nano Nomads

1. Cocucci, E., & Meldolesi, J. (2011). Ectosomes. *Current Biology*, 21(23), R940-R941.
2. Valadi, H., Ekström, K., Bossios, A., Sjöstrand, M., Lee, J. J., & Lötvall, J. O. (2007). Exosome-mediated transfer of mRNAs and microRNAs is a novel mechanism of genetic exchange between cells. *Nature Cell Biology*, 9(6), 654–659.
3. Maguire, G. (2016). Exosomes: Smart nanospheres for drug delivery naturally produced by stem cells. In A. M. Grumezescu (ed.), *Fabrication and Self-Assembly of Nanobiomaterials* (pp. 179–209). Oxford, UK: William Andrew Publishing.
4. Hoshino, A., Costa-Silva, B., Shen, T. L., Rodrigues, G., Hashimoto, A., Tesic Mark, M., ... & Lyden, D. (2015). Tumour exosome integrins determine organotropic metastasis. *Nature*, 527(7578), 329–335.
5. Andollo, N., Boyano, M. D., Andrade, R., Zalduendo, M. M., Eguizabal, C., Asumendi, A., ... & Arechaga, J. (2005). Structural and functional preservation of specific sequences of DNA and mRNA in apoptotic bodies from ES cells. *Apoptosis*, 10, 417–428.

6. Battistelli, M., & Falcieri, E. (2020). Apoptotic bodies: Particular extracellular vesicles involved in intercellular communication. *Biology*, 9(1), 21.
7. Krysko, D. V., Denecker, G., Festjens, N., Gabriels, S., Parthoens, E., D'Herde, K., & Vandenabeele, P. (2006). Macrophages use different internalization mechanisms to clear apoptotic and necrotic cells. *Cell Death & Differentiation*, 13(12), 2011–2022.
8. Schiller, M., Bekeredjian-Ding, I., Heyder, P., Blank, N., Ho, A. D., & Lorenz, H. M. (2008). Autoantigens are translocated into small apoptotic bodies during early stages of apoptosis. *Cell Death & Differentiation*, 15(1), 183–191.
9. Zhu, M., Zou, Q., Huang, R., Li, Y., Xing, X., Fang, J., ... & Yu, L. (2021). Lateral transfer of mRNA and protein by migrasomes modifies the recipient cells. *Cell Research*, 31(2), 237–240.
10. Zhang, Y., Guo, W., Bi, M., Liu, W., Zhou, L., Liu, H., ... & Xu, J. (2022). Migrasomes: From biogenesis, release, uptake, rupture to homeostasis and diseases. *Oxidative Medicine and Cellular Longevity*, 2022(1), 4525778.
11. Yu, S., & Yu, L. (2022). Migrasome biogenesis and functions. *The FEBS Journal*, 289(22), 7246–7254.
12. Tavano, S., & Heisenberg, C. P. (2019). Migrasomes take center stage. *Nature Cell Biology*, 21(8), 918–920.
13. Koupenova, M., Corkrey, H. A., Vitseva, O., Tanriverdi, K., Somasundaran, M., Liu, P., ... & Freedman, J. E. (2021). SARS-CoV-2 initiates programmed cell death in platelets. *Circulation Research*, 129(6), 631–646.
14. Meckes Jr, D. G., & Raab-Traub, N. (2011). Microvesicles and viral infection. *Journal of Virology*, 85(24), 12844–12854.
15. Valadi, H., Ekström, K., Bossios, A., Sjöstrand, M., Lee, J. J., & Lötvall, J. O. (2007). Exosome-mediated transfer of mRNAs and microRNAs is a novel mechanism of genetic exchange between cells. *Nature Cell Biology*, 9(6), 654–659.
16. Simeone, P., Bologna, G., Lanuti, P., Pierdomenico, L., Guagnano, M. T., Pieragostino, D., ... & Mariani-Costantini, R. (2020). Extracellular vesicles as signaling mediators and disease biomarkers across biological barriers. *International Journal of Molecular Sciences*, 21(7), 2514.
17. Clement, E., Lazar, I., Attané, C., Carrié, L., Dauvillier, S., Ducoux-Petit, M., ... & Nieto, L. (2020). Adipocyte extracellular vesicles carry enzymes and fatty acids that stimulate mitochondrial metabolism and remodeling in tumor cells. *The EMBO Journal*, 39(3), e102525.
18. Wang, S., He, B., Wu, H., Cai, Q., Ramírez-Sánchez, O., Abreu-Goodger, C., ... & Jin, H. (2024). Plant mRNAs move into a fungal pathogen via extracellular vesicles to reduce infection. *Cell Host & Microbe*, 32(1), 93–105.

Chapter 10: Invisible Instigators

1. Omuta, M. (1966). The microcephalic children of Hiroshima. *Japan Quarterly*, 13(3), 375.

2. Hiyama, A., Nohara, C., Kinjo, S., Taira, W., Gima, S., Tanahara, A., & Otaki, J. M. (2012). The biological impacts of the Fukushima nuclear accident on the pale grass blue butterfly. *Scientific Reports*, 2(1), 570.
3. Hinton, T. G., Alexakhin, R., Balonov, M., Gentner, N., Hendry, J., Prister, B., ... & Woodhead, D. (2007). Radiation-induced effects on plants and animals: Findings of the United Nations Chernobyl Forum. *Health Physics*, 93(5), 427–440.
4. Löbrich, M., Cooper, P. K., and Rydberg, M. (1996). Non-random distribution of DNA double-strand breaks induced by particle irradiation. *International Journal of Radiation Biology*, 70(5), 493–503.
5. Jin, Y., Yie, T. A., & Carothers, A. M. (1995). Non-random deletions at the dihydrofolate reductase locus of Chinese hamster ovary cells induced by α-particles simulating radon. *Carcinogenesis*, 16(8), 1981–1991.
6. Kovalchuk, O., Kovalchuk, I., Arkhipov, A., Hohn, B., & Dubrova, Y. E. (2003). Extremely complex pattern of microsatellite mutation in the germline of wheat exposed to the post-Chernobyl radioactive contamination. *Mutation Research / Fundamental and Molecular Mechanisms of Mutagenesis*, 525(1–2), 93–101.
7. Forster, L., Forster, P., Lutz-Bonengel, S., Willkomm, H., & Brinkmann, B. (2002). Natural radioactivity and human mitochondrial DNA mutations. *Proceedings of the National Academy of Sciences*, 99(21), 13950–13954.
8. Nowell, P. C. (2007). Discovery of the Philadelphia chromosome: A personal perspective. *The Journal of Clinical Investigation*, 117(8), 2033–2035.
9. Moynahan, M. E. (2002). The cancer connection: BRCA1 and BRCA2 tumor suppression in mice and humans. *Oncogene*, 21(58), 8994–9007.
10. Knudson Jr, A. G. (1971). Mutation and cancer: Statistical study of retinoblastoma. *Proceedings of the National Academy of Sciences*, 68(4), 820–823.
11. Hanahan, D., & Weinberg, R. A. (2000). The hallmarks of cancer. *Cell*, 100(1), 57–70.
12. Hanahan, D., & Weinberg, R. A. (2011). Hallmarks of cancer: The next generation. *Cell*, 144(5), 646–674.
13. Lawrence, M. S., Stojanov, P., Mermel, C. H., Robinson, J. T., Garraway, L. A., Golub, T. R., ... & Getz, G. (2014). Discovery and saturation analysis of cancer genes across 21 tumour types. *Nature*, 505(7484), 495–501.
14. Lawrence, M. S., Stojanov, P., Polak, P., Kryukov, G. V., Cibulskis, K., Sivachenko, A., ... & Getz, G. (2013). Mutational heterogeneity in cancer and the search for new cancer-associated genes. *Nature*, 499(7457), 214–218.
15. Gorbunova, V., Seluanov, A., Mao, Z., & Hine, C. (2007). Changes in DNA repair during aging. *Nucleic Acids Research*, 35(22), 7466–7474.
16. Roberts, S. A., Sterling, J., Thompson, C., Harris, S., Mav, D., Shah, R., ... & Gordenin, D. A. (2012). Clustered mutations in yeast

and in human cancers can arise from damaged long single-strand DNA regions. *Molecular Cell*, 46(4), 424–435.
17. Schuster-Böckler, B., & Lehner, B. (2012). Chromatin organization is a major influence on regional mutation rates in human cancer cells. *Nature*, 488(7412), 504–507.
18. Cuykendall, T. N., Rubin, M. A., & Khurana, E. (2017). Non-coding genetic variation in cancer. *Current Opinion in Systems Biology*, 1, 9–15.
19. Talavera, D., Taylor, M. S., & Thornton, J. M. (2010). The (non) malignancy of cancerous amino acidic substitutions. *Proteins: Structure, Function, and Bioinformatics*, 78(3), 518–529.
20. Li, S., & Wu, X. (2020). Common fragile sites: Protection and repair. *Cell & Bioscience*, 10(1), 29.
21. McLaughlin-Drubin, M. E., & Munger, K. (2008). Viruses associated with human cancer. *Biochimica et Biophysica Acta (BBA) – Molecular Basis of Disease*, 1782(3), 127–150.
22. Kost-Alimova, M., & Imreh, S. (2007, February). Modeling non-random deletions in cancer. *Seminars in Cancer Biology*, 17(1), 19–30.
23. McAvoy, S., Ganapathiraju, S. C., Ducharme-Smith, A. L., Pritchett, J. R., Kosari, F., Perez, D. S., ... & Smith, D. I. (2007). Non-random inactivation of large common fragile site genes in different cancers. *Cytogenetic and Genome Research*, 118(2–4), 260–269.

Chapter 11: Targeted Tethering

1. Liu, S. V., Nagasaka, M., Atz, J., Solca, F., & Müllauer, L. (2025). Oncogenic gene fusions in cancer: From biology to therapy. *Signal Transduction and Targeted Therapy*, 10(1), 111.
2. Tuna, M., Amos, C. I., & Mills, G. B. (2019). Molecular mechanisms and pathobiology of oncogenic fusion transcripts in epithelial tumors. *Oncotarget*, 10(21), 2095.
3. Shorokhova, M., Nikolsky, N., & Grinchuk, T. (2021). Chromothripsis – explosion in genetic science. *Cells*, 10(5), 1102.
4. Yanai, I., Wolf, Y. I., & Koonin, E. V. (2002). Evolution of gene fusions: Horizontal transfer versus independent events. *Genome Biology*, 3(5), research0024–1.
5. Snel, B., Bork, P., & Huynen, M. (2000). Genome evolution: Gene fusion versus gene fission. *Trends in Genetics*, 16(1), 9–11.
6. Nisole, S., Lynch, C., Stoye, J. P., & Yap, M. W. (2004). A Trim5-cyclophilin A fusion protein found in owl monkey kidney cells can restrict HIV-1. *Proceedings of the National Academy of Sciences*, 101(36), 13324–13328.
7. Virgen, C. A., Kratovac, Z., Bieniasz, P. D., & Hatziioannou, T. (2008). Independent genesis of chimeric TRIM5-cyclophilin proteins in two primate species. *Proceedings of the National Academy of Sciences*, 105(9), 3563–3568.
8. Yap, M. W., Dodding, M. P., & Stoye, J. P. (2006). Trim-cyclophilin A fusion proteins can restrict human immunodeficiency virus type 1

infection at two distinct phases in the viral life cycle. *Journal of Virology*, 80(8), 4061–4067.

9. Marcotte, E. M., Pellegrini, M., Ng, H. L., Rice, D. W., Yeates, T. O., & Eisenberg, D. (1999). Detecting protein function and protein-protein interactions from genome sequences. *Science*, 285(5428), 751–753.
10. Bolotin, E., Melamed, D., & Livnat, A. (2023). Genes that are used together are more likely to be fused together in evolution by mutational mechanisms: A bioinformatic test of the used-fused hypothesis. *Evolutionary Biology*, 50(1), 30–55.
11. Livnat, A. (2013). Interaction-based evolution: How natural selection and nonrandom mutation work together. *Biology Direct*, 8(1), 24.

Chapter 12: Reproductive Revolutions

1. Dudgeon, C. L., Coulton, L., Bone, R., Ovenden, J. R., & Thomas, S. (2017). Switch from sexual to parthenogenetic reproduction in a zebra shark. *Scientific Reports*, 7(1), 1–8.
2. Pamilo, P., Nei, M., & Li, W. H. (1987). Accumulation of mutations in sexual and asexual populations. *Genetics Research*, 49(2), 135–146.
3. Liu, W. S., Eriksson, L., & Fredga, K. (1998). XY sex reversal in the wood lemming is associated with deletion of Xp21–23 as revealed by chromosome microdissection and fluorescence in situ hybridization. *Chromosome Research*, 6, 379–384.
4. Song, S., & Zhang, J. (2024). In search of the genetic variants of human sex ratio at birth: Was Fisher wrong about sex ratio evolution? *Proceedings of the Royal Society B: Biological Sciences*, 291(2033), 20241876.
5. Haber, J. E. (1974). Bisexual mating behavior in a diploid of *Saccharomyces cerevisiae*: Evidence for genetically controlled non-random chromosome loss during vegetative growth. *Genetics*, 78(3), 843–858.
6. Smith, K. R. (2002). The role of sperm-mediated gene transfer in genome mutation and evolution. *Medical Hypotheses*, 59(4), 433–437.
7. Kaessmann, H. (2010). Origins, evolution, and phenotypic impact of new genes. *Genome Research*, 20(10), 1313–1326.
8. Drost, J. B., & Lee, W. R. (1995). Biological basis of germline mutation: Comparisons of spontaneous germline mutation rates among drosophila, mouse, and human. *Environmental and Molecular Mutagenesis*, 25(S2), 48–64.
9. Kong, A., Frigge, M. L., Masson, G., Besenbacher, S., Sulem, P., Magnusson, G., … & Stefansson, K. (2012). Rate of de novo mutations and the importance of father's age to disease risk. *Nature*, 488(7412), 471–475.
10. Wilson, J., Staley, J. M., & Wyckoff, G. J. (2020). Extinction of chromosomes due to specialization is a universal occurrence. *Scientific Reports*, 10(1), 2170.
11. McVean, G. T., & Hurst, L. D. (1997). Evidence for a selectively favourable reduction in the mutation rate of the X chromosome. *Nature*, 386(6623), 388–392.

12. Ellegren, H. (2011). Sex-chromosome evolution: Recent progress and the influence of male and female heterogamety. *Nature Reviews Genetics*, 12(3), 157–166.
13. Schartl, M., & Lamatsch, D. K. (2023). How to manage without a Y chromosome. *Proceedings of the National Academy of Sciences*, 120(2), e2218839120.
14. Blaser, O., Neuenschwander, S., & Perrin, N. (2014). Sex-chromosome turnovers: The hot-potato model. *The American Naturalist*, 183(1), 140–146.
15. Jeffries, D. L., Lavanchy, G., Sermier, R., Sredl, M. J., Miura, I., Borzée, A., ... & Perrin, N. (2018). A rapid rate of sex-chromosome turnover and non-random transitions in true frogs. *Nature Communications*, 9(1), 4088.

Chapter 13: Dizzying Duplications

1. Zeng, Q., Chen, H., Zhang, C., Han, M., Li, T., Qi, X., ... & He, N. (2015). Definition of eight mulberry species in the genus Morus by internal transcribed spacer-based phylogeny. *PloS One*, 10(8), e0135411.
2. Manimekalai, R., Suresh, G., Govinda Kurup, H., Athiappan, S., & Kandalam, M. (2020). Role of NGS and SNP genotyping methods in sugarcane improvement programs. *Critical Reviews in Biotechnology*, 40(6), 865–880.
3. Wendel, J. F. (2015). The wondrous cycles of polyploidy in plants. *American Journal of Botany*, 102(11).
4. Edger, P. P., & Pires, J. C. (2009). Gene and genome duplications: The impact of dosage-sensitivity on the fate of nuclear genes. *Chromosome Research*, 17, 699–717.
5. Dehal, P., & Boore, J. L. (2005). Two rounds of whole genome duplication in the ancestral vertebrate. *PLoS Biology*, 3(10), e314.
6. Van de Peer, Y., Mizrachi, E., & Marchal, K. (2017). The evolutionary significance of polyploidy. *Nature Reviews Genetics*, 18(7), 411–424.
7. Liu, C., Ren, Y., Li, Z., Hu, Q., Yin, L., Wang, H., ... & Fan, W. (2021). Giant African snail genomes provide insights into molluscan whole-genome duplication and aquatic–terrestrial transition. *Molecular Ecology Resources*, 21(2), 478–494.
8. Gui, Y., Yin, H., He, J. Y., Yang, S. H., Walsh, M. P., & Zheng, X. L. (2007). Endoreduplication of human smooth muscle cells induced by 2-methoxyestradiol: A role for cyclin-dependent kinase 2. *American Journal of Physiology – Heart and Circulatory Physiology*, 292(3), H1313–H1320. Schoenfelder, K. P., & Fox, D. T. (2015). The expanding implications of polyploidy. *The Journal of Cell Biology*, 209(4), 485.
9. Kolarčik, V., Frakova, V., Kocova, V., Koprivý, L., & Martonfi, P. (2020). Endopolyploidy pattern in Corydalis early spring geophytes. *Flora*, 270, 151651.
10. Zedek, F., Plačková, K., Veselý, P., Šmerda, J., Šmarda, P., Horová, L., & Bureš, P. (2020). Endopolyploidy is a common response to UV-B

stress in natural plant populations, but its magnitude may be affected by chromosome type. *Annals of Botany*, 126(5), 883–889.

11. Scholes, D. R., & Paige, K. N. (2015). Plasticity in ploidy: A generalized response to stress. *Trends in Plant Science*, 20(3), 165–175.
12. Paige, K. N. (2018). Overcompensation, environmental stress, and the role of endoreduplication. *American Journal of Botany*, 105(7), 1105–1108.
13. Neiman, M., Beaton, M. J., Hessen, D. O., Jeyasingh, P. D., & Weider, L. J. (2017). Endopolyploidy as a potential driver of animal ecology and evolution. *Biological Reviews*, 92(1), 234–247.
14. Losick, V. P., Fox, D. T., & Spradling, A. C. (2013). Polyploidization and cell fusion contribute to wound healing in the adult Drosophila epithelium. *Current Biology*, 23(22), 2224–2232.
15. Neiman, M., Beaton, M. J., Hessen, D. O., Jeyasingh, P. D., & Weider, L. J. (2017). Endopolyploidy as a potential driver of animal ecology and evolution. *Biological Reviews*, 92(1), 234–247.
16. Ramsey, J., & Schemske, D. W. (1998). Pathways, mechanisms, and rates of polyploid formation in flowering plants. *Annual Review of Ecology and Systematics*, 29(1), 467–501.
17. Van de Peer, Y., Mizrachi, E., & Marchal, K. (2017). The evolutionary significance of polyploidy. *Nature Reviews Genetics*, 18(7), 411–424.
18. Yao, Y., Carretero-Paulet, L., & Van de Peer, Y. (2019). Using digital organisms to study the evolutionary consequences of whole genome duplication and polyploidy. *PloS One*, 14(7), e0220257.
19. Yang, P. M., Huang, Q. C., Qin, G. Y., Zhao, S. P., & Zhou, J. G. (2014). Different drought-stress responses in photosynthesis and reactive oxygen metabolism between autotetraploid and diploid rice. *Photosynthetica*, 52(2), 193–202.
20. Hannweg, K., Steyn, W., & Bertling, I. (2016). In vitro-induced tetraploids of *Plectranthus esculentus* are nematode-tolerant and have enhanced nutritional value. *Euphytica*, 207(2), 343–351.
21. Godfree, R. C., Marshall, D. J., Young, A. G., Miller, C. H., & Mathews, S. (2017). Empirical evidence of fixed and homeostatic patterns of polyploid advantage in a keystone grass exposed to drought and heat stress. *Royal Society Open Science*, 4(11), 170934.
22. Mao, H., Chen, M., Su, Y., Wu, N., Yuan, M., Yuan, S., … & Chen, Y. (2018). Comparison on photosynthesis and antioxidant defense systems in wheat with different ploidy levels and octoploid triticale. *International Journal of Molecular Sciences*, 19(10), 3006.
23. Gunn, B. F., Murphy, D. J., Walsh, N. G., Conran, J. G., Pires, J. C., Macfarlane, T. D., & Birch, J. L. (2020). Evolution of Lomandroideae: Multiple origins of polyploidy and biome occupancy in Australia. *Molecular Phylogenetics and Evolution*, 149, 106836.
24. Rice, A., Šmarda, P., Novosolov, M., Drori, M., Glick, L., Sabath, N., … & Mayrose, I. (2019). The global biogeography of polyploid plants. *Nature Ecology & Evolution*, 3(2), 265–273.

25. Brochmann, C., Brysting, A. K., Alsos, I. G., Borgen, L., Grundt, H. H., Scheen, A. C., & Elven, R. (2004). Polyploidy in arctic plants. *Biological Journal of the Linnean Society*, 82(4), 521–536.
26. Van de Peer, Y., Mizrachi, E., & Marchal, K. (2017). The evolutionary significance of polyploidy. *Nature Reviews Genetics*, 18(7), 411–424.
27. Doyle, J. J., & Coate, J. E. (2019). Polyploidy, the nucleotype, and novelty: The impact of genome doubling on the biology of the cell. *International Journal of Plant Sciences*, 180(1), 1–52.
28. Freeling, M. (2017). Picking up the ball at the K/Pg boundary: The distribution of ancient polyploidies in the plant phylogenetic tree as a spandrel of asexuality with occasional sex. *The Plant Cell*, 29(2), 202–206.
29. Bretagnolle, F. A., & Thompson, J. D. (1995). Gametes with the somatic chromosome number: Mechanisms of their formation and role in the evolution of autopolyploid plants. *New Phytologist*, 129(1), 1–22.
30. Mable, B. K., Alexandrou, M. A., & Taylor, M. I. (2011). Genome duplication in amphibians and fish: An extended synthesis. *Journal of Zoology*, 284(3), 151–182.
31. Ramsey, J., & Schemske, D. W. (1998). Pathways, mechanisms, and rates of polyploid formation in flowering plants. *Annual Review of Ecology and Systematics*, 29(1), 467–501.
32. Kürschner, W. M., Batenburg, S. J., & Mander, L. (2013). Aberrant Cl-assopollis pollen reveals evidence for unreduced (2 n) pollen in the conifer family Cheirolepidiaceae during the Triassic–Jurassic transition. *Proceedings of the Royal Society B: Biological Sciences*, 280(1768), 20131708.
33. Raven, P. H. (1975). The bases of angiosperm phylogeny: Cytology. *Annals of the Missouri Botanical Garden*, 62, 724–764.
34. Soltis, D. E., Soltis, P. S., Bennett, M. D., & Leitch, I. J. (2003). Evolution of genome size in the angiosperms. *American Journal of Botany*, 90(11), 1596–1603.
35. Wolfe, K. (2000). Robustness – it's not where you think it is. *Nature Genetics*, 25(1), 3–4.
36. Scannell, D. R., Frank, A. C., Conant, G. C., Byrne, K. P., Woolfit, M., & Wolfe, K. H. (2007). Independent sorting-out of thousands of duplicated gene pairs in two yeast species descended from a whole-genome duplication. *Proceedings of the National Academy of Sciences*, 104(20), 8397–8402.
37. Byrne, K. P., & Wolfe, K. H. (2005). The Yeast Gene Order Browser: Combining curated homology and syntenic context reveals gene fate in polyploid species. *Genome Research*, 15(10), 1456–1461.
38. Conant, G. C. (2014). Comparative genomics as a time machine: How relative gene dosage and metabolic requirements shaped the time-dependent resolution of yeast polyploidy. *Molecular Biology and Evolution*, 31(12), 3184–3193.
39. Emery, M., Willis, M. M. S., Hao, Y., Barry, K., Oakgrove, K., Peng, Y., ... & Conant, G. C. (2018). Preferential retention of genes from

one parental genome after polyploidy illustrates the nature and scope of the genomic conflicts induced by hybridization. *PLoS Genetics*, 14(3), e1007267.

40. De Smet, R., Adams, K. L., Vandepoele, K., Van Montagu, M. C., Maere, S., & Van de Peer, Y. (2013). Convergent gene loss following gene and genome duplications creates single-copy families in flowering plants. *Proceedings of the National Academy of Sciences*, 110(8), 2898–2903.
41. Emery, M., Willis, M. M. S., Hao, Y., Barry, K., Oakgrove, K., Peng, Y., ... & Conant, G. C. (2018). Preferential retention of genes from one parental genome after polyploidy illustrates the nature and scope of the genomic conflicts induced by hybridization. *PLoS Genetics*, 14(3), e1007267.
42. Renny-Byfield, S., Gong, L., Gallagher, J. P., & Wendel, J. F. (2015). Persistence of subgenomes in paleopolyploid cotton after 60 My of evolution. *Molecular Biology and Evolution*, 32(4), 1063–1071.
43. Emery, M., Willis, M. M. S., Hao, Y., Barry, K., Oakgrove, K., Peng, Y., ... & Conant, G. C. (2018). Preferential retention of genes from one parental genome after polyploidy illustrates the nature and scope of the genomic conflicts induced by hybridization. *PLoS Genetics*, 14(3), e1007267.

Chapter 14: Compensatory Coding

1. Dufresne, F. R. A. N. C. E. (2011). The history of the Daphnia pulex complex: Asexuality, hybridization, and polyploidy. *Phylogeography and Population Genetics in Crustacea*, 217–232.
2. Hartl, D. L., & Taubes, C. H. (1996). Compensatory nearly neutral mutations: Selection without adaptation. *Journal of Theoretical Biology*, 182(3), 303–309.
3. Lang, G. I., Rice, D. P., Hickman, M. J., Sodergren, E., Weinstock, G. M., Botstein, D., & Desai, M. M. (2013). Pervasive genetic hitchhiking and clonal interference in forty evolving yeast populations. *Nature*, 500(7464), 571–574.
4. Harcombe, W. R., Springman, R., & Bull, J. J. (2009). Compensatory evolution for a gene deletion is not limited to its immediate functional network. *BMC Evolutionary Biology*, 9, 1–11.
5. Andersson, D. I., & Hughes, D. (2010). Antibiotic resistance and its cost: Is it possible to reverse resistance? *Nature Reviews Microbiology*, 8(4), 260–271.
6. Barešić, A., Hopcroft, L. E., Rogers, H. H., Hurst, J. M., & Martin, A. C. (2010). Compensated pathogenic deviations: Analysis of structural effects. *Journal of Molecular Biology*, 396(1), 19–30.
7. Poon, A., & Chao, L. (2005). The rate of compensatory mutation in the DNA bacteriophage φX174. *Genetics*, 170(3), 989–999.
8. Rokyta, D., Badgett, M. R., Molineux, I. J., & Bull, J. J. (2002). Experimental genomic evolution: Extensive compensation for loss of

DNA ligase activity in a virus. *Molecular Biology and Evolution*, 19(3), 230–238.
9. Burch, C. L., & Chao, L. (1999). Evolution by small steps and rugged landscapes in the RNA virus ϕ6. *Genetics*, 151(3), 921–927.
10. Szamecz, B., Boross, G., Kalapis, D., Kovács, K., Fekete, G., Farkas, Z., … & Pál, C. (2014). The genomic landscape of compensatory evolution. *PLoS Biology*, 12(8), e1001935.
11. San Millan, A., & MacLean, R. C. (2017). Fitness costs of plasmids: A limit to plasmid transmission. *Microbiology Spectrum*, 5(5), MTBP-0016–2017.
12. Zwanzig, M., Harrison, E., Brockhurst, M. A., Hall, J. P., Berendonk, T. U., & Berger, U. (2019). Mobile compensatory mutations promote plasmid survival. *mSystems*, 4(1), e00186–18.
13. Comas, I., Borrell, S., Roetzer, A., Rose, G., Malla, B., Kato-Maeda, M., … & Gagneux, S. (2012). Whole-genome sequencing of rifampicin-resistant Mycobacterium tuberculosis strains identifies compensatory mutations in RNA polymerase genes. *Nature Genetics*, 44(1), 106–110.
14. Davis, B. H., Poon, A. F., & Whitlock, M. C. (2009). Compensatory mutations are repeatable and clustered within proteins. *Proceedings of the Royal Society B: Biological Sciences*, 276(1663), 1823–1827.

Chapter 15: Arcane Aggregation

1. Korsten, G., Pelle, R., Hoogenberg, B., Kampinga, H. H., & Kapitein, L. C. (2022). Nuclear poly-glutamine aggregates rupture the nuclear envelope and hinder its repair. *bioRxiv*, 2022–11.
2. Chiti, F., & Dobson, C. M. (2006). Protein misfolding, functional amyloid, and human disease. *Annual Review of Biochemistry*, 75(1), 333–366.
3. Lindner, A. B., Madden, R., Demarez, A., Stewart, E. J., & Taddei, F. (2008). Asymmetric segregation of protein aggregates is associated with cellular aging and rejuvenation. *Proceedings of the National Academy of Sciences*, 105(8), 3076–3081.
4. Rolli, S., Langridge, C. A., & Sontag, E. M. (2024). Clearing the JUNQ: The molecular machinery for sequestration, localization, and degradation of the JUNQ compartment. *Frontiers in Molecular Biosciences*, 11, 1427542.
5. Rothe, S., Prakash, A., & Tyedmers, J. (2018). The insoluble protein deposit (IPOD) in yeast. *Frontiers in Molecular Neuroscience*, 11, 237.
6. Johnston, J. A., Ward, C. L., & Kopito, R. R. (1998). Aggresomes: A cellular response to misfolded proteins. *The Journal of Cell Biology*, 143(7), 1883–1898.
7. Dobson, C. M. (2003). Protein folding and misfolding. *Nature*, 426(6968), 884–890.
8. Plakoutsi, G., Bemporad, F., Calamai, M., Taddei, N., Dobson, C. M., & Chiti, F. (2005). Evidence for a mechanism of amyloid formation involving molecular reorganisation within native-like precursor aggregates. *Journal of Molecular Biology*, 351(4), 910–922.

9. Meisl, G., Xu, C. K., Taylor, J. D., Michaels, T. C., Levin, A., Otzen, D., … & Knowles, T. P. (2022). Uncovering the universality of self-replication in protein aggregation and its link to disease. *Science Advances*, 8(31), eabn6831.
10. Shmookler Reis, R. J., Atluri, R., Balasubramaniam, M., Johnson, J., Ganne, A., & Ayyadevara, S. (2021). "Protein aggregates" contain RNA and DNA, entrapped by misfolded proteins but largely rescued by slowing translational elongation. *Aging Cell*, 20(5), e13326.
11. Bernardi, L., & Bruni, A. C. (2019). Mutations in prion protein gene: Pathogenic mechanisms in C-terminal vs. N-terminal domain, a review. *International Journal of Molecular Sciences*, 20(14), 3606.
12. Appleby, B. S., Shetty, S., & Elkasaby, M. (2022). Genetic aspects of human prion diseases. *Frontiers in Neurology*, 13, 1003056.
13. Tartaglia, G. G., Pechmann, S., Dobson, C. M., & Vendruscolo, M. (2007). Life on the edge: A link between gene expression levels and aggregation rates of human proteins. *Trends in Biochemical Sciences*, 32(5), 204–206.
14. Drummond, D. A., Bloom, J. D., Adami, C., Wilke, C. O., & Arnold, F. H. (2005). Why highly expressed proteins evolve slowly. *Proceedings of the National Academy of Sciences*, 102(40), 14338–14343.
15. De Baets, G., Reumers, J., Delgado Blanco, J., Dopazo, J., Schymkowitz, J., & Rousseau, F. (2011). An evolutionary trade-off between protein turnover rate and protein aggregation favors a higher aggregation propensity in fast degrading proteins. *PLoS Computational Biology*, 7(6), e1002090.
16. Chen, Y., & Dokholyan, N. V. (2008). Natural selection against protein aggregation on self-interacting and essential proteins in yeast, fly, and worm. *Molecular Biology and Evolution*, 25(8).
17. Tartaglia, G. G., Pechmann, S., Dobson, C. M., & Vendruscolo, M. (2007). Life on the edge: A link between gene expression levels and aggregation rates of human proteins. *Trends in Biochemical Sciences*, 32(5), 204–206.
18. Chiti, F., & Dobson, C. M. (2006). Protein misfolding, functional amyloid, and human disease. *Annual Review of Biochemistry*, 75(1), 333–366.
19. Rousseau, F., Serrano, L., & Schymkowitz, J. W. (2006). How evolutionary pressure against protein aggregation shaped chaperone specificity. *Journal of Molecular Biology*, 355(5), 1037–1047.
20. Rauscher, S., Baud, S., Miao, M., Keeley, F. W., & Pomes, R. (2006). Proline and glycine control protein self-organization into elastomeric or amyloid fibrils. *Structure*, 14(11), 1667–1676.
21. Schwartz, R., Istrail, S., & King, J. (2001). Frequencies of amino acid strings in globular protein sequences indicate suppression of blocks of consecutive hydrophobic residues. *Protein Science*, 10(5), 1023–1031.
22. Foy, S. G., Wilson, B. A., Bertram, J., Cordes, M. H., & Masel, J. (2019). A shift in aggregation avoidance strategy marks a long-term direction to protein evolution. *Genetics*, 211(4), 1345–1355.

23. Monsellier, E., & Chiti, F. (2007). Prevention of amyloid-like aggregation as a driving force of protein evolution. *EMBO Reports*, 8(8), 737–742.

Chapter 16: Genetic Genesis

1. Schlötterer, C. (2015). Genes from scratch – the evolutionary fate of de novo genes. *Trends in Genetics*, 31(4), 215–219.
2. Dujon, B. (1996). The yeast genome project: What did we learn? *Trends in Genetics*, 12(7), 263–270.
3. Zhao, L., Saelao, P., Jones, C. D., & Begun, D. J. (2014). Origin and spread of de novo genes in Drosophila melanogaster populations. *Science*, 343(6172), 769–772.
4. Toll-Riera, M., Bosch, N., Bellora, N., Castelo, R., Armengol, L., Estivill, X., & Mar Alba, M. (2009). Origin of primate orphan genes: A comparative genomics approach. *Molecular Biology and Evolution*, 26(3), 603–612.
5. McLysaght, A., & Hurst, L. D. (2016). Open questions in the study of de novo genes: What, how and why. *Nature Reviews Genetics*, 17(9), 567–578.
6. Levine, M. T., Jones, C. D., Kern, A. D., Lindfors, H. A., & Begun, D. J. (2006). Novel genes derived from noncoding DNA in Drosophila melanogaster are frequently X-linked and exhibit testis-biased expression. *Proceedings of the National Academy of Sciences*, 103(26), 9935–9939.
7. Donoghue, M. T., Keshavaiah, C., Swamidatta, S. H., & Spillane, C. (2011). Evolutionary origins of Brassicaceae specific genes in *Arabidopsis thaliana. BMC Evolutionary Biology*, 11, 1–23.
8. Schlotterer, C. (2015). Genes from scratch.
9. Khalturin, K., Hemmrich, G., Fraune, S., Augustin, R., & Bosch, T. C. (2009). More than just orphans: Are taxonomically-restricted genes important in evolution? *Trends in Genetics*, 25(9), 404–413.
10. Clark, M. B., Amaral, P. P., Schlesinger, F. J., Dinger, M. E., Taft, R. J., Rinn, J. L., ... & Mattick, J. S. (2011). The reality of pervasive transcription. *PLoS Biology*, 9(7), e1000625.
11. Neme, R., & Tautz, D. (2016). Fast turnover of genome transcription across evolutionary time exposes entire non-coding DNA to de novo gene emergence. *eLife*, 5, e09977.
12. Schmidt, E. E. (1996). Transcriptional promiscuity in testes. *Current Biology*, 6(7), 768–769.
13. Wu, D. D., Irwin, D. M., & Zhang, Y. P. (2011). De novo origin of human protein-coding genes. *PLoS Genetics*, 7(11), e1002379.
14. Ingolia, N. T., Brar, G. A., Stern-Ginossar, N., Harris, M. S., Talhouarne, G. J., Jackson, S. E., ... & Weissman, J. S. (2014). Ribosome profiling reveals pervasive translation outside of annotated protein-coding genes. *Cell Reports*, 8(5), 1365–1379.
15. Tretyachenko, V., Vymětal, J., Bednárová, L., Kopecký Jr, V., Hofbauerová, K., Jindrová, H., ... & Hlouchová, K. (2017). Random protein sequences can form defined secondary structures and are well-tolerated in vivo. *Scientific Reports*, 7(1), 15449.

16. Kosinski, L. J., Aviles, N. R., Gomez, K., & Masel, J. (2022). Random peptides rich in small and disorder-promoting amino acids are less likely to be harmful. *Genome Biology and Evolution*, 14(6), evac085.
17. Weisman, C. M. (2022). The origins and functions of de novo genes: Against all odds? *Journal of Molecular Evolution*, 90(3), 244–257.
18. Tautz, D., & Domazet-Lošo, T. (2011). The evolutionary origin of orphan genes. *Nature Reviews Genetics*, 12(10), 692–702.
19. Palmieri, N., Kosiol, C., & Schlötterer, C. (2014). The life cycle of Drosophila orphan genes. *eLife*, 3, e01311.
20. Anand, A., Olson, C. A., Yang, L., Sastry, A. V., Catoiu, E., Choudhary, K. S., … & Palsson, B. O. (2019). Pseudogene repair driven by selection pressure applied in experimental evolution. *Nature Microbiology*, 4(3), 386–389.
21. Zheng, D., & Gerstein, M. B. (2007). The ambiguous boundary between genes and pseudogenes: The dead rise up, or do they? *Trends in Genetics*, 23(5), 219–224.
22. Vihinen, M. (2014). Contribution of pseudogenes to sequence diversity. In L. Poliseno (ed.), *Pseudogenes: Functions and Protocols*, (pp. 15–24). New York: Humana Press.
23. Dewannieux, M., Esnault, C., & Heidmann, T. (2003). LINE-mediated retrotransposition of marked Alu sequences. *Nature Genetics*, 35(1), 41–48.
24. Zheng, D., Frankish, A., Baertsch, R., Kapranov, P., Reymond, A., Choo, S. W., … & Gerstein, M. B. (2007). Pseudogenes in the ENCODE regions: Consensus annotation, analysis of transcription, and evolution. *Genome Research*, 17(6), 839–851.
25. Pei, B., Sisu, C., Frankish, A., Howald, C., Habegger, L., Mu, X. J., … & Gerstein, M. B. (2012). The GENCODE pseudogene resource. *Genome Biology*, 13, 1–26.
26. Prade, V. M., Gundlach, H., Twardziok, S., Chapman, B., Tan, C., Langridge, P., … & Mayer, K. F. (2018). The pseudogenes of barley. *The Plant Journal*, 93(3), 502–514.
27. Podlaha, O., & Zhang, J. (2010, 15 November). Pseudogenes and their evolution. *Encyclopedia of Life Sciences (ELS)*, 1–8.
28. Inai, Y., Ohta, Y., & Nishikimi, M. (2003). The whole structure of the human nonfunctional L-gulono-gamma-lactone oxidase gene – the gene responsible for scurvy – and the evolution of repetitive sequences thereon. *Journal of Nutritional Science and Vitaminology*, 49(5), 315–319.

Epilogue

1. Webb, C. J., & Zakian, V. A. (2016). Telomerase RNA is more than a DNA template. *RNA Biology*, 13(8), 683–689.
2. Johnston, W. K., Unrau, P. J., Lawrence, M. S., Glasner, M. E., & Bartel, D. P. (2001). RNA-catalyzed RNA polymerization: Accurate and general RNA-templated primer extension. *Science*, 292(5520), 1319–1325.

3. Gilbert, W. (1986). Origin of life: The RNA world. *Nature*, 319(6055), 618.
4. Adkar-Purushothama, C. R., & Perreault, J. P. (2020). Impact of nucleic acid sequencing on viroid biology. *International Journal of Molecular Sciences*, 21(15), 5532.
5. Rout, S. K., Wunnava, S., Krepl, M., Cassone, G., Šponer, J. E., Mast, C. B., ... & Braun, D. (2025). Amino acids catalyse RNA formation under ambient alkaline conditions. *Nature Communications*, 16(1), 1–9.
6. Müller, F., Escobar, L., Xu, F., Węgrzyn, E., Nainytė, M., Amatov, T., ... & Carell, T. (2022). A prebiotically plausible scenario of an RNA–peptide world. *Nature*, 605(7909), 279–284.
7. Hud, N. V. (2018). Searching for lost nucleotides of the pre-RNA World with a self-refining model of early Earth. *Nature Communications*, 9(1), 5171.
8. Diemer, G. S., & Stedman, K. M. (2012). A novel virus genome discovered in an extreme environment suggests recombination between unrelated groups of RNA and DNA viruses. *Biology Direct*, 7, 1–14.
9. Ehrenfreund, P., Rasmussen, S., Cleaves, J., & Chen, L. (2006). Experimentally tracing the key steps in the origin of life: The aromatic world. *Astrobiology*, 6(3), 490–520.
10. Tielens, A. G. G. M. (2013). The molecular universe. *Reviews of Modern Physics*, 85(3), 1021–1081.
11. Becker, L., Glavin, D. P., & Bada, J. L. (1997). Polycyclic aromatic hydrocarbons (PAHs) in Antarctic Martian meteorites, carbonaceous chondrites, and polar ice. *Geochimica et Cosmochimica Acta*, 61(2), 475–481.
12. Beintema, D. A. (1997). PAH's and crystalline silicates in planetary nebulae. *Astrophysics and Space Science*, 255, 507–512.

Index

www.ingramcontent.com/pod-product-compliance
Ingram Content Group UK Ltd.
Pitfield, Milton Keynes, MK11 3LW, UK
UKHW012013200826
474373UK00001BA/1

9 781784 276836